Existence and
Stability of
Nash Equilibrium

Existence and Stability of Nash Equilibrium

Guilherme Carmona

University of Cambridge, UK &
Universidade Nova de Lisboa, Portugal

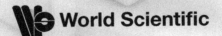
World Scientific

NEW JERSEY • LONDON • SINGAPORE • BEIJING • SHANGHAI • HONG KONG • TAIPEI • CHENNAI

Published by

World Scientific Publishing Co. Pte. Ltd.

5 Toh Tuck Link, Singapore 596224

USA office: 27 Warren Street, Suite 401-402, Hackensack, NJ 07601

UK office: 57 Shelton Street, Covent Garden, London WC2H 9HE

Library of Congress Cataloging-in-Publication Data
Carmona, Guilherme.
  Existence and stability of Nash equilibrium / by Guilherme Carmona.
    p. cm.
  Includes bibliographical references and index.
  ISBN 978-9814390651
  1. Game theory. 2. Equilibrium (Economics)   I. Title.
  HB144.C367 2013
  519.3--dc23
                            2012031330

British Library Cataloguing-in-Publication Data
A catalogue record for this book is available from the British Library.

In-house Editor: Alisha Nguyen

Typeset by Stallion Press
Email: enquiries@stalliuonpress.com

Printed in Singapore.

For Filipa, Manuel and Carlota

Preface

The question of existence of Nash equilibrium has a beautiful history that has been enriched recently through several developments. The purpose of this book is to present such developments and to clarify the relationship between several of them.

This book is largely based on my own work as an author, referee and editor. It, therefore, reflects my taste and my understanding of the problem of existence of Nash equilibrium. Nevertheless, I hope it can be a useful tools for those who wish to learn about this topic, to apply the results presented here or to extend them in new directions.

Acknowledgements

Much of my work on existence of equilibrium has been done together with Konrad Podczeck, and it is a pleasure to acknowledge his contribution to this book. My understanding of this problem was also greatly enhanced with conversations with, and therefore I thank, Adib Bagh, Erik Balder, Paulo Barelli, Mehmet Barlo, Luciano de Castro, Partha Dasgupta, José Fajardo, Andy McLennan, Phil Reny, Hamid Sabourian and Nicholas Yannelis. I also thank Alisha Nguyen, the editor of this book at World Scientific Publishing, for her efficiency. Financial support from Fundação para a Ciência e a Tecnologia is gratefully acknowledged.

Contents

Chapter 1

Introduction

Game theory studies situations characterized by strategic interaction of several individuals, arising when the well-being of at least one individual depends on what other individuals may do. The analysis of such situations takes two steps. First, they are formally described by a game, which, when represented in normal form, is a list of players (the individuals), strategies (what these individuals can choose), and payoffs (representing the well-being of individuals associated with different strategies). Second, to the game describing the situation being analyzed, a solution concept is applied to identify strategies with special properties. The interpretation of the resulting equilibrium strategies is that they form reasonable predictions for the problem being studied. Thus, the existence of an equilibrium means that there is at least one reasonable prediction for the problem in hand, and this justifies the importance that the question of existence of equilibrium has historically received.

Nash equilibrium is, perhaps, the most widely used solution concept for normal-form games. It was originally defined in Nash (1950), where its existence was established under general conditions. These conditions include, in particular, the continuity of players' payoff functions. Chapter 2 reviews results on the existence of Nash equilibrium in continuous games and also introduces the stability problem we consider. Of the many possible stability problems one can consider, we focus on the question of how the set of Nash equilibria of a normal-form games changes with changes in the elements defining the game (namely, the players' strategy spaces and payoff functions). Chapter 2 also presents some stability results for the case of continuous games.

Motivated by some economic problems that are naturally modeled by games with discontinuous payoff functions, Dasgupta and Maskin (1986a),

Simon (1987), Simon and Zame (1990), Lebrun (1996), Reny (1999), Carmona (2009), Barelli and Soza (2010), McLennan, Monteiro and Tourky (2011), among others, have considerably extended Nash's result and technique (see also Carmona, 2011b, on which this introduction is based). This book presents several advances to the literature on existence of Nash equilibrium for games with discontinuous payoff functions that emerged from those papers.

We start by considering (generalized) better-reply secure games [introduced by Reny (1999) and generalized by Barelli and Soza (2010)]. We will show that generalized better-reply secure games have Nash equilibria and that the existence of Nash equilibrium in these games can be understood in light of the following simple observation: A Nash equilibrium for a game with compact strategy sets exists if (1) the game can be suitable replaced by a better-behaved of game, (2) the better-behaved game has a sequence of approximate equilibria with a vanishing level of approximation, and (3) all limit points of every sequence of approximate equilibria, with a level of approximation converging to zero, of the better-behaved game is a Nash equilibrium of the original game.

The existence result for generalized better-reply secure games and the approach on which it is based are then extended to yield stronger existence results in Chapter 4.

The above approach to address the question of existence of Nash equilibrium in generalized better-reply secure games, in particular, its third step illustrates the importance that limit results have for that question. Limit results are also central to the stability problem addressed in this book. We present several limit results in Chapter 5 and show their implications for existence and stability of Nash equilibria.

Limit results are also useful to address the existence and stability of solutions of the games with an endogenous sharing rules introduced by Simon and Zame (1990). Such results are presented in Chapter 6.

Another setting where the existence and stability results for normal-form games are useful is that of games with a continuum of players, as described in Schmeidler (1973) and Mas-Colell (1984) among others. Games with a continuum of players with finite characteristics (i.e., finite action spaces and payoff functions) are considered in Chapter 7. In this chapter, we present a notion of generalized better-reply security for games with a continuum of players and show that all such games have an equilibrium distribution.

Chapter 2

Continuous Normal-Form Games

The main topics of this book — existence and stability of Nash equilibria — are now well understood in games with continuous payoff functions. The results obtained for this case are reviewed in this chapter, as well as the standard techniques used to derive them. This chapter also introduces the notation and some key concepts that will be used throughout this book.

We start by showing the existence of pure strategy Nash equilibria in games where each player's payoff function is (jointly) continuous and quasiconcave in the player's own action. This result is establishing by applying a fixed point theorem to the game's best-reply correspondence, an argument that goes back to Nash (1950).

We then consider continuous games that fail to be quasiconcave (in the above sense). Nevertheless, such games have Nash equilibria in mixed strategies. This conclusion requires defining the mixed extension of a game as the game where players are allowed to randomize between the actions originally available to them and to compute payoffs in the mixed extension as the expected value of payoffs in the original game. The pure strategy Nash equilibria of the mixed extension of a game are, by definition, the mixed strategy equilibria of the original game and their existence follows from the existence of pure strategy Nash equilibria in continuous quasiconcave games.

The notion of stability we focus on concerns how the set of Nash equilibria changes when we change players' payoff functions. This issue is analyzed by studying the Nash equilibrium correspondence, which maps games into their set of Nash equilibria. We show that, in the space of continuous and quasiconcave games, the Nash equilibrium correspondence is upper hemicontinuous and that the set of games where it fails to be continuous is exceptional.

The result showing that the Nash equilibrium correspondence is upper hemicontinuous is a limit result in the following sense. It considers a sequence of games whose payoff functions converge uniformly to the payoff function of a limit game and a corresponding sequence of Nash equilibria converging to some limit strategy. It then shows that the limit strategy is a Nash equilibrium of the limit game.

Limit results such as the above are also useful to establish the existence of Nash equilibria. In particular, we show that a slightly different limit result, together with existence of approximate equilibria in a given game, imply the existence of Nash equilibria in that game.

2.1 Notation and Definitions

We are mostly concerned with games in normal form. A game in normal form is defined by listing the players in the game, their actions and their preferences (represented by payoff functions) over the possible action profiles. Thus, a *normal-form game* $G = (X_i, u_i)_{i \in N}$ consists of a finite set of players $N = \{1, \ldots, n\}$ and, for all players $i \in N$, a strategy space X_i and a payoff function $u_i : X \to \mathbb{R}$, where $X = \prod_{i \in N} X_i$.

The solution concept for normal-form games that we focus on is Nash equilibrium. The formal definition of this concept is as follows. Let $G = (X_i, u_i)_{i \in N}$ be a normal-form game and $i \in N$. The symbol $-i$ denotes "all players but i" and $X_{-i} = \prod_{j \neq i} X_j$ denotes the set of strategy profiles for all players but i. A Nash equilibrium is a strategy profile with the property that the strategy of each player is an optimal choice given the strategy of the other players. Thus, a strategy $x^* \in X$ is a *Nash equilibrium of G* if $u_i(x^*) \geq u_i(x_i, x^*_{-i})$ for all $i \in N$ and $x_i \in X_i$. We let $E(G)$ denote the set of Nash equilibria of G.

The optimality criterion in the definition of Nash equilibrium can be written using the concept of a player's value function. Formally, *player i's value function* is the function $w_{u_i} : X_{-i} \to \mathbb{R}$ defined by $w_{u_i}(x_{-i}) = \sup_{x_i \in X_i} u_i(x_i, x_{-i})$ for all $x_{-i} \in X_{-i}$. We sometimes use v_i instead of w_{u_i} to denote player i's value function. Loosely, $v_i(x_{-i})$ is the highest possible payoff that player i can obtain when the other players choose x_{-i}. Thus, equivalently, a strategy $x^* \in X$ is Nash equilibrium if

$$u_i(x^*) \geq v_i(x^*_{-i}) \quad \text{for all } i \in N. \tag{2.1}$$

We also consider a weaker solution concept which is useful for addressing the existence of equilibrium in the discontinuous normal-form games.

Such weaker solution concept is obtained by replacing, in (2.1), each player's value function with a function strictly below it. Intuitively, such function represents, for each player, a less demanding aspiration level than his value function. Let $G = (X_i, u_i)_{i \in N}$ be a normal-form game and define $F(G)$ to be the set of all functions $f = (f_1, \ldots, f_n)$ such that f_i is a real-valued function on X and $f_i(x) \leq v_i(x_{-i})$ for all $x \in X$ and $i \in N$. For all $f \in F(G)$, we say a strategy $x^* \in X$ is an f-*equilibrium of G* if $u_i(x^*) \geq f_i(x^*)$ for all $i \in N$. It is clear that f-equilibrium is a weaker solution concept that Nash equilibrium in the sense that if $x^* \in X$ is a Nash equilibrium of G then x^* is a f-equilibrium for all $f \in F(G)$. We also note that the concept of ε-equilibrium, where $\varepsilon > 0$, is a particular case of f-equilibrium. In fact, for all $\varepsilon > 0$, x^* is an ε-*equilibrium* if it is an f-equilibrium for $f = (v_1 - \varepsilon, \ldots, v_n - \varepsilon)$.

We classify normal-form games according to the properties of their action spaces and payoff functions. We say that a normal-form game G is: (1) *metric* if X_i is a metric space for all $i \in N$; (2) *compact* if X_i is compact and u_i is bounded for all $i \in N$; (3) *quasiconcave* if, for all $i \in N$, X_i is a convex subset of a topological vector space and $u_i(\cdot, x_{-i})$ is quasiconcave for all $x_{-i} \in X_{-i}$; and (4) *continuous* if u_i is continuous for all $i \in N$.

We focus on compact metric games and we let $\mathbb{G}$ denote the class of such games. The set of games in $\mathbb{G}$ that are quasiconcave plays an important role and is denoted by $\mathbb{G}_q$.

2.2   Existence of Nash Equilibria

Existence of Nash equilibrium in continuous, compact and quasiconcave games is well-know since the work of Nash (1950). The key idea of Nash's argument is that the set of Nash equilibria of a normal-form game equals the set of fixed points of the game's best-reply correspondence.

The best-reply correspondence of a game maps each strategy profile x into the set of strategies profiles y with the property that each player's action in y maximizes her payoff function given that the others are playing according to x. Thus, a fixed point of the better-reply correspondence, which is a strategy profile that belongs to the set of best-replies to itself, is a Nash equilibrium. Therefore, it follows that a Nash equilibrium exists whenever the best-reply correspondence has a fixed point. The assumptions of continuity, compactness and quasiconcavity are then made to ensure that the best-reply correspondence satisfies a set of sufficient conditions for the existence of a fixed point.

Using the above reasoning, in the simpler case where the game in question is also metric, we obtain the following basic existence result for normal-form games.

Theorem 2.1   *If $G = (X_i, u_i)_{i \in N} \in \mathbb{G}_q$ is continuous, then G has a Nash equilibrium.*

As mentioned above, the proof of Theorem 2.1 is based on a fixed-point argument using the best-reply correspondence of a game. The *best-reply correspondence* of a normal-form game $G = (X_i, u_i)_{i \in N}$, denoted by $B : X \rightrightarrows X$, is defined by setting

$$B(x) = \{y \in X : u_i(y_i, x_{-i}) = v_i(x_{-i}) \text{ for all } i \in N\}$$

for all $x \in X$.

Lemma 2.2 establishes some properties of the best-reply correspondence of games that are metric, compact, quasiconcave and continuous. Some definitions are needed.

Let Y and Z be metric spaces and $\Psi : Y \rightrightarrows Z$ be a correspondence. We say that Ψ is *upper hemicontinuous* if, for all $y \in Y$ and all open $U \subseteq Z$ such that $\Psi(y) \subseteq U$, there exists a neighborhood V of y such that $\Psi(y') \subseteq U$ for all $y' \in V$. Furthermore, Ψ has *nonempty (resp. convex, closed) values* if $\Psi(y)$ is nonempty (resp. convex, closed) for all $y \in Y$. To simplify the terminology, we say that Ψ is *well-behaved* if Ψ is upper hemicontinuous with nonempty and closed values. In the case where Z is also a vector space, then we abuse terminology and say that Ψ is well-behaved if, besides being upper hemicontinuous with nonempty closed values, Ψ is also convex-valued.

Lemma 2.2   *If $G = (X_i, u_i)_{i \in N} \in \mathbb{G}_q$ is continuous, then the best-reply correspondence of G is well-behaved.*

Proof.   Let $x \in X$ and $i \in N$. Since X_i is compact and u_i is continuous, it follows from Theorem A.11 that there exists $y_i \in X_i$ such that $u_i(y_i, x_{-i}) = v_i(x_{-i})$. Hence, it follows that $y = (y_1, \dots, y_n)$ belongs to $B(x)$, which implies that B has nonempty values.

Let $y, y' \in B(x)$, $\lambda \in (0, 1)$ and $i \in N$. Since $u_i(\cdot, x_{-i})$ is quasiconcave, one obtains that $u_i(\lambda y_i + (1 - \lambda)y_i', x_{-i}) \geq \min\{u_i(y_i, x_{-i}), u_i(y_i', x_{-i})\}$. Thus, $u_i(\lambda y_i + (1 - \lambda)y_i', x_{-i}) = v_i(x_{-i})$. Hence, $\lambda y + (1 - \lambda)y' = (\lambda y_1 + (1 - \lambda)y_1', \dots, \lambda y_n + (1 - \lambda)y_n') \in B(x)$ and B has convex values.

Let $\{(x_k, y_k)\}_{k=1}^{\infty} \subseteq \text{graph}(B)$ be a convergent sequence, $(x, y) = \lim_k (x_k, y_k)$ and $i \in N$. Then, $u_i(y_i^k, x_{-i}^k) = v_i(x_{-i}^k)$ for all $k \in \mathbb{N}$, and so the continuity of u_i and v_i (the latter follows from Theorems A.12 and A.13) implies that $u_i(y_i, x_{-i}) = \lim_k u_i(y_i^k, x_{-i}^k) = \lim_k v_i(x_{-i}^k) = v_i(x_{-i})$. Since $u_i(y_i, x_{-i}) = v_i(x_{-i})$ for all $i \in N$, it follows that $y \in B(x)$ and so $(x, y) \in \text{graph}(B)$. Thus, $\text{graph}(B)$ is closed and, by Theorem A.5, B is upper hemicontinuous and has closed values. ∎

The properties of the best-reply correspondence described in Lemma 2.2 imply that the Cauty's fixed-point theorem (Theorem A.14) applies. This yields a fixed point for the best-reply correspondence and, consequently, a Nash equilibrium for the normal-form game.

Proof of Theorem 2.1.   It follows from Theorem A.14 and Lemma 2.2 that B has a fixed point, i.e., there exists $x^* \in X$ such that $x^* \in B(x^*)$. Hence, $u_i(x^*) = v_i(x_{-i}^*)$ for all $i \in N$ and, therefore, x^* is a Nash equilibrium of G. ∎

2.3 Mixed Strategies

An important property for the existence of Nash equilibrium as stated in Theorem 2.1 is the quasiconcavity of the game. The importance of this property is easily illustrated by the matching pennies game. This is a two-player game ($N = \{1, 2\}$) in which each player's action space is $X_1 = X_2 = \{H, T\}$, where H stands for "heads" and T for "tails." Player 1 wins player 2's penny if the players' choices match and player 2 wins player 1's penny if the players' choices do not match. Hence, players' payoff function are $u_1(x_1, x_2) = 1$ and $u_2(x_1, x_2) = -1$ if $x_1 = x_2$ and, if $x_1 \neq x_2$, $u_1(x_1, x_2) = -1$ and $u_2(x_1, x_2) = 1$. It is easy to see that this game has no Nash equilibrium and that all the assumptions of Theorem 2.1 are satisfied but quasiconcavity.

The standard way to deal with the above problem is to introduce mixed strategies. This amounts to say that each player can randomize over the actions in his action space. For example, some player may choose one action with probability 1/3 and another action with probability 2/3.

The usefulness of introducing mixed strategies is that it transforms any non-quasiconcave game in a quasiconcave game. The new game is called the mixed extension of the original game and is such that each player's strategy space is the set of probability measures over the set of strategies of the original game, the latter now referred to as the set of pure strategies. This choice for the strategy space of the mixed extension of a game already implies that one of the requirements of quasiconcavity is satisfied, namely the convexity of the strategy space. In fact, the set of probability measure of the set of strategies of the original game is a convex set: For example, if σ_i and σ_i' are two mixed strategies of player i that, for simplicity, assign a non-zero probability only to two pure strategies x and x', say $\sigma_i(\hat{x})$ and $\sigma_i'(\hat{x})$ for all $\hat{x} \in \{x, x'\}$, then, for all $\lambda \in (0, 1)$, $\lambda\sigma_i + (1 - \lambda)\sigma_i'$ is a mixed strategy for player i that assigns probability $\lambda\sigma_i(x) + (1 - \lambda)\sigma_i'(x)$ to x and $\lambda\sigma_i(x') + (1 - \lambda)\sigma_i'(x')$ to x'.

Furthermore, the second requirement of quasiconcavity, namely the quasi-concavity of each player's payoff function in his own strategy is also satisfied in the mixed extension of any normal-form game. This is obtained by assuming that each player's payoff of a profile of mixed strategies is the expected value of that player's payoff function with respect to the joint probability distribution over pure strategy profiles.

The formal definition of mixed strategies and of the mixed extension of a normal-form game is as follows. Let $G = (X_i, u_i)_{i \in N}$ be a compact metric normal-form game in which u_i is measurable for all $i \in N$. For all $i \in N$, let M_i denote the set of Borel probability measures on X_i. Players are assumed to randomize independently. Thus, the probability over profiles of pure strategies induced by a profile of mixed strategies $m = (m_1, \ldots, m_n)$ is the product measure $m_1 \times \cdots \times m_n$. For all $i \in N$, player i's payoff function in the mixed extension of G is $u_i : M \to \mathbb{R}$ defined by $u_i(m) = \int_X u_i(x) \mathrm{d}(m_1 \times \cdots \times m_n)(x)$ for all $m = (m_1, \ldots, m_n) \in M = \prod_{i \in N} M_i$. The *mixed extension* of a normal-form game $G = (X_i, u_i)_{i \in N}$ is the normal-form game $\overline{G} = (M_i, u_i)_{i \in N}$.

The mixed extension of a normal-form game is itself a normal-form game and, therefore, the definition of Nash equilibrium applies. Each of the (pure strategy) Nash equilibria of the mixed extension $\overline{G}$ of a normal-form game $G = (X_i, u_i)_{i \in N}$ is called a *mixed strategy Nash equilibrium* of G.

The reason for considering the mixed extension $\overline{G}$ of a normal-form game G is, as discussed above, due to the fact that $\overline{G}$ is quasiconcave even when G is not. Theorem 2.3 below shows that the remaining properties we mentioned in the previous section, metrizability, compactness and continuity, are inherited by the mixed extension of any normal-form that possesses them. Since these properties are topological, their discussion requires us to endow the mixed strategy sets with a topology.

Given a normal-form game $G = (X_i, u_i)_{i \in N}$ and its mixed extension $\overline{G} = (M_i, \bar{u}_i)_{i \in N}$, we endow M_i, for all $i \in N$, with the narrow topology. The narrow topology on M_i is the coarsest topology on M_i that makes the map $\mu \mapsto \int_{X_i} f \mathrm{d}\mu$ continuous for every bounded continuous real-valued function f on X_i.

Theorem 2.3   *If $G = (X_i, u_i)_{i \in N}$ is a metric, compact and continuous normal-form game, then $\overline{G} = (M_i, u_i)_{i \in N}$ is metric, compact, continuous and quasiconcave.*

Proof.   It follows from Theorem A.21 that M_i is compact and metric for all $i \in N$. Since $x \mapsto u_i(x)$ is bounded, it follows that $m \mapsto u_i(m)$ is bounded for all $i \in N$. Hence, $\overline{G}$ is metric and compact.

By Theorem A.24, it follows that $m \mapsto m_1 \times \cdots \times m_n$ is continuous. Since $x \mapsto u_i(x)$ is continuous, this implies that $m \mapsto u_i(m)$ is continuous for all $i \in N$. Hence, $\overline{G}$ is continuous.

It is clear that M_i is convex and that

$$u_i\left(\lambda m_i + (1 - \lambda)m_i', m_{-i}\right) = \lambda u_i(m_i, m_{-i}) + (1 - \lambda)u_i(m_i', m_{-i}) \quad (2.2)$$

for all $i \in N$, $m_i, m_i' \in M_i$ and $m_{-i} \in M_{-i}$. This implies that $\overline{G}$ is quasiconcave. ■

For any normal-form game G, its mixed extension $\overline{G}$ is, by Theorem 2.3, a metric, compact, quasiconcave and continuous normal-form game. Hence, Theorem 2.1 implies that $\overline{G}$ has a pure strategy Nash equilibrium. Thus, G has a mixed strategy Nash equilibrium.

Theorem 2.4   *If $G = (X_i, u_i)_{i \in N} \in \mathbb{G}$ is continuous, then G has a mixed strategy Nash equilibrium.*

2.4   Stability of Nash Equilibria

The notion of stability we consider in this book concerns how the set of Nash equilibria changes with changes in the elements defining the game. Recall that a normal-form game is defined by the set of players, the players' strategy spaces and players' payoff function; of these elements, we focus on players' payoff functions. Thus, the question we address can be phased as follows: When will two games with payoff functions that are close to each other have sets of Nash equilibria that are also closed to each other?

To address the above question, we define the notion of the Nash equilibrium correspondence, a notion of distance between payoff functions and, thus, a notion of distance between games.

Let N be a finite set of players and $(X_i)_{i \in N}$ be a collection of compact, convex subsets of a metric vector space. Let, as before, $X = \prod_{i \in N} X_i$ and let $C(X)$ denote the space of real-valued continuous functions on X. Furthermore, let $A(X) \subset C(X)^n$ be the space of continuous functions $u = (u_1, \ldots, u_n) : X \to \mathbb{R}^n$ such that $u_i(\cdot, x_{-i})$ is quasiconcave for all $i \in N$ and $x_{-i} \in X_{-i}$. Let $\mathbb{G}_c(X)$ be the set of normal-form games $G = (X_i, u_i)_{i \in N}$ such that $u = (u_1, \ldots, u_n) \in A(X)$. We make $\mathbb{G}_c(X)$ a metric space by defining $d(G, G') = \max_{i \in N} \sup_{x \in X} |u_i(x) - u_i'(x)|$ for all $G, G' \in \mathbb{G}_c(X)$. Thus, a sequence $\{G_k\}_{k=1}^{\infty} \subseteq \mathbb{G}_c(X)$ converges to G if and only if $\{u_k\}_{k=1}^{\infty} \subseteq A(X)$ converges to u uniformly.

The Nash equilibrium correspondence is $E : \mathbb{G}_c(X) \rightrightarrows X$ defined by

$$E(G) = \{x \in X : x \text{ is a Nash equilibrium of } G\}$$

for all $G \in \mathbb{G}_c(X)$.

Given the above definition, we can restate our initial questions as asking when will the Nash equilibrium correspondence be continuous. We have introduced previously a notion of continuity of correspondences, namely that of upper hemicontinuity. Another notion of continuity of correspondences is the following one. Let Y and Z be metric spaces and $\Psi : Y \rightrightarrows Z$ be a correspondence. We say that Ψ is *lower hemicontinuous at* $y \in Y$ if, for all open $U \subseteq Z$ such that $\Psi(y) \cap U \neq \emptyset$, there exists a neighborhood V of y such that $\Psi(y') \cap U \neq \emptyset$ for all $y' \in V$. If Ψ is lower hemicontinuous at y for all $y \in Y$, then we say that Ψ is *lower hemicontinuous*. Combining the two notions of continuity, we say that Ψ is *continuous at* $y \in Y$ if Ψ is both upper and lower hemicontinuous at y; furthermore, Ψ is *continuous* if Ψ is continuous at y for all $y \in Y$.

Before addressing the continuity of the equilibrium correspondence, we start by identifying sufficient conditions for the limit points of sequences of approximate equilibria of games converging to a limit game to be Nash equilibria of the limit game.

Theorem 2.5   *Let* $G \in \mathbb{G}_c(X)$, $\{G_k\}_{k=1}^{\infty} \subseteq \mathbb{G}_c(X)$, $\{f_k\}_{k=1}^{\infty}$ *be such that* $f_k \in F(G_k)$ *for all* $k \in \mathbb{N}$, $\{x_k\}_{k=1}^{\infty} \subseteq X$ *and* $x \in X$. *If* $G = \lim_k G_k$, $x = \lim_k x_k$, x_k *is* f_k-*equilibrium of* G_k *for all* $k \in \mathbb{N}$ *and* $\liminf_k f_i^k(x_k) \geq v_i(x_{-i})$ *for all* $i \in N$, *then* x *is Nash equilibrium of* G.

Proof.   Let $i \in N$. Then,

$$u_i(x) = \lim_k u_i^k(x_k) \geq \liminf_k f_i^k(x_k) \geq v_i(x_{-i}).$$

Since $i \in N$ is arbitrary, it follows that x is a Nash equilibrium of G.   ∎

An implication of Theorem 2.5 is the upper hemicontinuity of the Nash equilibrium correspondence.

Theorem 2.6   *The Nash equilibrium correspondence E is upper hemicontinuous.*

Proof.   Due to Theorem A.5, it suffices to show that E is closed. Let $(G, x) \in \mathbb{G}_c(X) \times X$ and $\{G_k, x_k\}_{k=1}^{\infty} \subseteq \text{graph}(E)$ be such that $\lim_k (G_k, x_k) = (G, x)$.

Given $u \in A(X)$, let $\|u\| = \max_{i \in N} \sup_{x \in X} |u_i(x)|$. Let $u, \tilde{u} \in A(X)$, $v = w_u$ and $\tilde{v} = w_{\tilde{u}}$. Then $\|v - \tilde{v}\| \leq \|u - \tilde{u}\|$. In fact, for all $i \in N$, $x_{-i} \in X_{-i}$ and $\varepsilon > 0$, there is some $x_i \in X_i$ such that $v_i(x_{-i}) < u_i(x_i, x_{-i}) - \varepsilon \leq \tilde{u}_i(x_i, x_{-i}) + \|u - \tilde{u}\| - \varepsilon \leq \tilde{v}_i(x_{-i}) + \|u - \tilde{u}\| - \varepsilon$. Since $\varepsilon > 0$ is arbitrary, then $v_i(x_{-i}) \leq \tilde{v}_i(x_{-i}) + \|u - \tilde{u}\|$. Reversing the role of v_i and $\tilde{v}_i$, we get $\tilde{v}_i(x_{-i}) \leq v_i(x_{-i}) + \|u - \tilde{u}\|$ and, hence, $\|v - \tilde{v}\| \leq \|u - \tilde{u}\|$.

It follows from the above argument that $\{v_k\}_k$ converges uniformly to v, where $v_k = w_{u_k}$ for all $k \in \mathbb{N}$. Hence, $\lim_k v_i^k(x_k) = v_i(x)$ for all $i \in N$. It then follows from Theorem 2.5 that $x \in E(G)$. Thus, E is closed. ■

Although the Nash equilibrium correspondence is upper hemicontinuous in the domain of continuous games, it is not lower hemicontinuous. This is shown by the following example. For all $k \in \mathbb{N}$, let G_k be the mixed extension of the finite normal form game defined by Table 2.1.

We have that $\{G_k\}_{k=1}^{\infty}$ converges to the game G defined as the mixed extension of the normal-form game with payoff function given by $u_i(x) = 0$ for all $i \in \{1, 2\}$ and $x \in \{A, B\}^2$. Hence, $E(G) = M(X)$. In contrast, $E(G_k) = \{(A, A)\}$ for all $k \in \mathbb{N}$.

The above example is simple because it relies on an extreme property of the limit game G: each player in G is indifferent between all action profiles. While it shows easily the failure of the to be lower hemicontinuity equilibrium correspondence, even in the domain of continuous games, the example and the property on which it relies is, intuitively, rather exceptional. We will show next that, in a formal sense, this property and, more generally,

Table 2.1.   Payoffs for G_k.

1\2	A	B
A	$\frac{1}{k}, \frac{1}{k}$	$\frac{1}{k}, 0$
B	$0, \frac{1}{k}$	$0, 0$

the failure of the equilibrium correspondence to be lower hemicontinuity is exceptional.

We start with the special case considered in the above examples, where, for all $i \in N$, there exists a finite set F_i such that $X_i = M(F_i)$, i.e. each player's strategy space is the set of mixed strategies over a finite set of pure strategies. In this case, each player's payoff function is fully described by a vector in $\mathbb{R}^{m'}$ where $m' = \prod_{i \in N} |F_i|$ is the number of action profiles. Thus, when $X = \prod_{i \in N} X_i$ is being held fixed, a game is fully described by players' payoff function which can be regarded as a vector in $\mathbb{R}^{nm'}$. Letting $m = nm'$, this means, in particular, that we can regard the Nash equilibrium correspondence as a correspondence from $\mathbb{R}^m$ to X. Using this special feature, the next result shows that the set of games at which the Nash equilibrium correspondence fails to be continuous is contained in a closed set of Lebesgue measure zero.

Theorem 2.7 *Suppose that $X_i = M(F_i)$ with F_i finite for all $i \in N$ and let*

$$C = \{u \in \mathbb{R}^m : E \text{ is continuous at } u\}.$$

Then $\mathrm{cl}(C^c)$ *has Lebesgue measure zero.*

Proof. Let G be such a game. We say that $x \in E(G)$ is essential if for all $\varepsilon > 0$ there is $\delta > 0$ such that for all $G' \in \mathbb{G}$ such that $d(G, G') < \delta$ there exists $x' \in E(G')$ with $d(x, x') < \varepsilon$.

Note that E is lower hemicontinuous at G if and only if all Nash equilibria of G are essential. Since the complement of the set of games with the property that all its Nash equilibria is essential has a null closure [see Theorem 2.6.2 in van Damme (1991) and also Harsanyi (1973)], it follows that $\mathrm{cl}(C^c)$ has Lebesgue measure zero. ∎

Theorem 2.7 provides a formal sense according to which the failure of the Nash equilibrium correspondence to be continuous is exceptional. It relies on the existence of a particular measure on the space of games with appealing properties and whose existence relies on the special nature of the problem, i.e., on the fact that players' strategy spaces are probability measures over a finite set of pure actions.

The next result describes a property of the set C defined in Theorem 2.7 that is easily applied to the general case of metric strategy spaces. Let S be a metric space. A subset $T \subseteq S$ is *nowhere dense* if $\mathrm{int}(\mathrm{cl}(T)) = \emptyset$, it is *first category in S* if T is a countable union of nowhere dense sets and is *second category in S* if T is not first category.

Theorem 2.8 *Suppose that $X_i = M(F_i)$ with F_i finite for all $i \in N$ and let*

$$C = \{u \in \mathbb{R}^m : E \text{ is continuous at } u\}.$$

Then C is second category in $\mathbb{R}^m$.

Proof. Let λ denote the Lebesgue measure in $\mathbb{R}^m$. Since, by Theorem 2.7, $\lambda(\mathrm{cl}(C^c)) = 0$ and $\lambda(O) > 0$ for all nonempty open sets $O \in \mathbb{R}^m$, it follows that $\mathrm{int}(\mathrm{cl}(C^c)) = \emptyset$. Thus, C^c is nowhere dense and, in particular, first category in $\mathbb{R}^m$.

Note that the union of two first category sets is also a first category set. Furthermore, by Theorem A.10, $\mathbb{R}^m$ is second category in itself. Thus, C is second category in $\mathbb{R}^m$; otherwise, $\mathbb{R}^m = C \cup C^c$ would be first category in itself. ∎

The conclusion of Theorem 2.8 will be used as our definition of a generic property in the general case of metric strategy spaces. With this definition, we obtain that the Nash equilibrium correspondence is, in the domain of continuous games, continuous except at an exceptional set of games.

Theorem 2.9 *Let $C = \{G \in \mathbb{G}_c(X) : E \text{ is continuous at } G\}$. Then C is second category in $\mathbb{G}_c(X)$.*

Proof. We have that E is upper hemicontinuous by Theorem 2.6. Thus, by Theorem A.9, C^c is first category in $\mathbb{G}_c(X)$.

Note that $\mathbb{G}_c(X)$ is a complete metric space. Indeed, if $\{G_k\}_{k=1}^{\infty}$ is a Cauchy sequence in $\mathbb{G}_c(X)$, then $\{u_k\}_{k=1}^{\infty}$ is a Cauchy sequence in $C(X)^n$. Since the latter space is complete, there exists $u \in C(X)^n$ such that $\lim_k u_k = u$. Thus, $G = (X_i, u_i)_{i \in N}$ is such that u is continuous. Furthermore, $u_i(\cdot, x_{-i})$ is quasiconcave for all $i \in N$ and $x_{-i} \in X_{-i}$ since, for all $\alpha \in \mathbb{R}$, $\{x_i \in X_i : u_i(x_i, x_{-i}) \geq \alpha\} = \cap_{k=1}^{\infty}\{x_i \in X_i : u_i^k(x_i, x_{-i}) \geq \alpha - \|u_k - u\|\}$ and $\{x_i \in X_i : u_i^k(x_i, x_{-i}) \geq \alpha - \|u_k - u\|\}$ is convex for all $k \in \mathbb{N}$. In conclusion, $G \in \mathbb{G}_c(X)$ and, therefore, $\mathbb{G}_c(X)$ is complete.

It then follows by Theorem A.10 that $\mathbb{G}_c(X)$ is second category in itself. This, together with the fact that C^c is first category in $\mathbb{G}_c(X)$, implies that C is second category in $\mathbb{G}_c(X)$. ∎

2.5 Existence of Nash Equilibria via Approximate Equilibria

The results we have established so far — existence of equilibrium via fixed points and limit results — can be combined to give an alternative proof of the existence result for continuous quasiconcave games (Theorem 2.1). Namely, one can first establish the existence of approximate equilibria using a fixed point argument and then use a limit result to conclude that any limit point of a sequence of approximate equilibria, with a level of approximation converging to zero, is a Nash equilibrium.

Although the above approach is unnecessarily complicated in the case of continuous games, it turns out to be useful in the class of discontinuous games considered in Chapter 3 which, in particular, may fail to have a well-behaved best-reply correspondence.

Another advantage of the above approach is that it allows us to establish existence of equilibrium in games where each player's strategy space is not locally convex without using Cauty's fixed point theorem. Instead, by first proving the existence of ε-equilibrium for all $\varepsilon > 0$ and then using the limit result provided in Theorem 2.5, the existence of equilibrium can be established using Browder's fixed point theorem.

The formal argument is as follows. Suppose that $G \in \mathbb{G}_q$ is continuous. Let $\varepsilon > 0$ and let $\Psi : X \rightrightarrows X$ be defined by $\Psi(x) = \{y \in X : u_i(y_i, x_{-i}) > v_i(x_{-i}) - \varepsilon$ for all $i \in N\}$ for all $x \in X$. For all $y \in X$, we have that $\Psi^{-1}(y) = \{x \in X : y \in \Psi(x)\}$ is open due to the continuity of both u_i and v_i for all $i \in N$. Furthermore, for all $x \in X$, $\Psi(x)$ is nonempty (by the definition of v_i) and convex (since G is quasiconcave). Hence, it follows by Theorem A.15 that Ψ has a fixed point and, thus, G has an ε-equilibrium.

Due to the above, for all $k \in \mathbb{N}$, let x_k be an $1/k$-equilibrium of G and let $f_k \in F(G)$ be defined by $f_i^k(x) = v_i(x) - 1/k$ for all $x \in X$ and $i \in N$. Since X is compact, we may assume, taking a subsequence if necessary, that $\{x_k\}_{k=1}^{\infty}$ converges. Let $x = \lim_k x_k$. The definition of $\{f_k\}_{k=1}^{\infty}$ and the continuity of v_i imply that $\liminf_k f_i^k(x_k) = v_i(x)$ for all $i \in N$. Hence, by Theorem 2.5, x is a Nash equilibrium of G.

The alternative proof of Theorem 2.1 raises the question of whether or not one can prove Cauty's fixed point theorem (at least in the special case of metric spaces) using the existence of equilibrium theorem for continuous games and, in particular, as a consequence of Browder's fixed point theorem. The answer to this questions seems to be negative. The reason is that, given a nonempty, convex, compact subset X of a metric space and a continuous function $f : X \to X$, the local convexity of the underlying metric space seems to be needed to establish the quasiconcavity of any game whose Nash equilibria are fixed points of f.

Although we do not prove that all continuous, metric games G whose Nash equilibria are fixed points of f are quasiconcave if and only if the underlying metric space is locally convex, the following discussion illustrates why local convexity is important. This will be done by using the local convexity of the underlying metric space to construct a quasiconcave, continuous, metric game whose Nash equilibria are fixed points of f.

Suppose that T is a metric space with metric d and that, for all $t \in T$ and $\varepsilon > 0$, $B_\varepsilon(t) = \{t' \in T : d(t,t') < \varepsilon\}$ is convex; hence, T is locally convex and, furthermore, $\bar{B}_\varepsilon(t) = \{t' \in T : d(t,t') \le \varepsilon\}$ is also convex for all $t \in T$ and $\varepsilon > 0$. Let $X \subseteq T$ be nonempty, convex and compact and $f : X \to X$ be continuous.

Define the following two-player game G as follows: $N = \{1,2\}$, $X_1 = X_2 = X$, $u_1(x_1,x_2) = -d(x_1,x_2)$ and $u_2(x_1,x_2) = -d(f(x_1),x_2)$ for all $(x_1,x_2) \in X_1 \times X_2$. It is clear that G is continuous and the convexity of $\bar{B}_\varepsilon(t) = \{t' \in T : d(t,t') \le \varepsilon\}$ for all $t \in T$ and $\varepsilon > 0$ implies that G is also quasiconcave. Hence, G has a Nash equilibrium (x_1^*, x_2^*) by Theorem 2.1, which clearly satisfies $x_1^* = x_2^*$ and $x_2^* = f(x_1^*)$. Thus, x_1^* is a fixed point of f.

Chapter 3

Generalized Better-Reply Secure Games

Many games of interest are naturally modeled with discontinuous payoff functions, several examples of those being presented in Dasgupta and Maskin (1986b). As a result, the results in Chapter 2 have been extended to a broader class of games including, in particular, games with discontinuous payoff functions. This line of research was initiated in Dasgupta and Maskin (1986a) and, later on, their results, as well as some others that followed, were unified by Reny (1999) through the notion of better-reply secure games.

In this chapter, we consider a generalization of better-reply security due to Barelli and Soza (2010). We first present an existence result for generalized better-reply secure games and illustrate it with two examples. The proof of this existence result is based on Carmona (2011c) and uses an argument similar to the one presented in Section 2.5. More precisely, we obtain first a sequence of approximate equilibria via a fixed point argument and, second, we apply a limit result to this sequence of approximate equilibria to obtain a Nash equilibrium of the game in question.

Generalized better-reply security can be understood, in part, as replacing players' payoff functions of a given game with a given better-behaved payoff function which is, in a precise sense, related to the original one. We consider the possibility of replacing the original payoff function with other functions and, also, to change the way the better-behaved function relates with the original one. This allows us to obtain two characterizations of generalized better-reply security that clarify the nature of the better-behaved payoff function and its relationship with the original payoff function implicit in the definition of generalized better-reply security. We also present analogous results for the case of better-reply security.

We then provide several sufficient conditions for a game to be generalized better-reply secure. Moreover, we establish conditions on the pure strategies of a game that ensure that its mixed extension is generalized better-reply secure.

3.1 Generalized Better-Reply Security and Existence of Equilibrium

The main existence result in this chapter (Theorem 3.2 below) states that all compact and quasiconcave games satisfying generalized better-reply security have a Nash equilibrium.

Generalized better-reply security guarantees that any game satisfying it can be approximated by a well-behaved game (in the sense that a fixed point argument can be used to establish the existence of approximate equilibria; Lemmas 3.4 and 3.5) and that the approximation can be done in such a way that limit points of approximate equilibria of the approximating game, with the level of approximation suitably converging to zero, are themselves Nash equilibria of the original game (Lemma 3.6).

The above argument is analogous to the one used to established the existence of Nash equilibrium in continuous games via approximate equilibria used in Section 2.5. The difference in generalized better-reply secure games is that, first, the fixed point argument is applied to a "regularized" game and, second, the notion of approximate equilibria needs to be more general than that of ε-equilibria.

The formal development of these ideas is as follows. Let $G = (X_i, u_i)_{i \in N}$ be a normal-form game and Γ be the closure of the graph of $u = (u_1, \ldots, u_n)$. We say that G is *generalized better-reply secure* if whenever $(x^*, u^*) \in \Gamma$ and x^* is not a Nash equilibrium, there exists a player $i \in N$, an open neighborhood U of x^*_{-i}, a well-behaved correspondence $\varphi_i : U \rightrightarrows X_i$, and a number $\alpha_i > u_i^*$ such that $u_i(x') \geq \alpha_i$ for all $x' \in \mathrm{graph}(\varphi_i)$.

Recall that the convention introduced in Chapter 2 according to which the meaning a well-behaved correspondence depends on whether its range space is a vector space or not. In particular, the above definition gives two notions of generalized better-reply security: one for games in $\mathbb{G}$, where the correspondence φ_i is required to be upper hemicontinuous, nonempty and closed-valued, and another for games in $\mathbb{G}_q$, where, in addition, φ_i

is required to be convex-valued. Here we make the convention that a statement made for a generalized better-reply secure game $G \in \mathbb{G}_q$ means that G is generalized better-reply secure in the latter sense, where a statement made for a generalized better-reply secure game $G \in \mathbb{G}$ means that G is generalized better-reply secure in the former sense.

The following result shows that the two notions of generalized better-reply security coincide for games in $\mathbb{G}_q \subseteq \mathbb{G}$ when the strategy spaces are subsets of a locally convex vector space.

Theorem 3.1 *Let $G = (X_i, u_i)_{i \in N} \in \mathbb{G}_q$ be such that X_i is a subset of a locally convex vector space. Then G is generalized better-reply secure in $\mathbb{G}_q$ if and only if G is generalized better-reply secure in $\mathbb{G}$.*

Proof.   Let G satisfy the above assumptions. It is clear that if G is generalized better-reply secure in $\mathbb{G}_q$ then G is generalized better-reply secure in $\mathbb{G}$.

Conversely, suppose that G is generalized better-reply secure in $\mathbb{G}$. Let $(x^*, u^*) \in \Gamma$ be such that x^* is not a Nash equilibrium. Since G is generalized better-reply secure in $\mathbb{G}$, there exists $i \in N$, an open neighborhood U of x^*_{-i}, a upper hemicontinuous correspondence $\tilde{\varphi}_i : U \rightrightarrows X_i$ with nonempty, closed (and, hence, compact) values, and $\alpha_i > u_i^*$ such that $u_i(x') \geq \alpha_i$ for all $x' \in \text{graph}(\tilde{\varphi}_i)$.

Define $\varphi_i : U \rightrightarrows X_i$ by $\varphi_i(x_{-i}) = \text{co}(\tilde{\varphi}_i(x_{-i}))$ for all $x_{-i} \in U$. Then, it follows by Theorem A.6 that φ_i is upper hemicontinuous with nonempty, convex and compact (hence, closed) values. Thus, G is generalized better-reply secure in $\mathbb{G}_q$. ∎

The following result establishes the existence of Nash equilibria in generalized better-reply secure games.

Theorem 3.2 *If $G = (X_i, u_i)_{i \in N} \in \mathbb{G}_q$ is generalized better-reply secure, then G has a Nash equilibrium.*

Theorem 3.2 is established with the help of three lemmas. The first of these lemmas describes the properties of the game $\underline{G}$ defined by changing the players' payoff functions as follows: For all $i \in N$ and $x_{-i} \in X_{-i}$, let $N(x_{-i})$ denote the set of all open neighborhoods of x_{-i}. Furthermore, for all $i \in N$, $x \in X$ and $U \in N(x_{-i})$, let $W_U(x)$ be the set of all well-behaved correspondences $\varphi_i : U \rightrightarrows X_i$ that satisfy $x \in \text{graph}(\varphi_i)$. For all $i \in N$ and

$x \in X$, define

$$\underline{u}_i(x) = \sup_{U \in N(x_{-i})} \sup_{\varphi_i \in W_U(x)} \inf_{z \in \text{graph}(\varphi_i)} u_i(z).$$

Finally, let $\underline{G} = (X_i, \underline{u}_i)_{i \in N}$.

Before we describe the properties of $\underline{G}$, we note that the game $\underline{G}$ is, through its players' value function, implicit in the definition of generalized better-reply security. This is shown in Theorem 3.3 below which provides a simple characterization of generalized better-reply security. In order to simplify the notation, we let, for all $i \in N$, $\underline{v}_i$ denote player i's value function in $\underline{G}$, i.e., $\underline{v}_i(x_{-i}) = \sup_{x_i \in X_i} \underline{u}_i(x_i, x_{-i})$ for all $x_{-i} \in X_{-i}$.

Theorem 3.3   *Let $G = (X_i, u_i)_{i \in N} \in \mathbb{G}_q$. Then, G is generalized better-reply secure if and only if x^* is a Nash equilibrium of G for all $(x^*, u^*) \in \Gamma$ such that $u_i^* \geq \underline{v}_i(x^*_{-i})$ for all $i \in N$.*

Proof.   (Necessity) Let $G \in \mathbb{G}_q$ be such that x^* is a Nash equilibrium of G for all $(x^*, u^*) \in \Gamma$ such that $u_i^* \geq \underline{v}_i(x^*_{-i})$ for all $i \in N$. Let $(x^*, u^*) \in \Gamma$ be such that x^* is not a Nash equilibrium. Thus, there is $i \in N$ such that $\underline{v}_i(x^*_{-i}) > u_i^*$. Hence, there exists $\alpha_i > u_i^*$ and $x_i \in X_i$ such that $\underline{u}_i(x_i, x^*_{-i}) > \alpha_i$. This, in turn, implies that there exist $U \in N(x^*_{-i})$ and $\varphi_i \in W_U(x_i, x^*_{-i})$ such that $u_i(z) > \alpha_i$ for all $z \in \text{graph}(\varphi_i)$. Thus, G is generalized better-reply secure.

(Sufficiency) Suppose that $G \in \mathbb{G}_q$ is generalized better-reply secure. Consider $(x^*, u^*) \in \Gamma$ such that $u_i^* \geq \underline{v}_i(x^*_{-i})$ for all $i \in N$, and, in order to reach a contradiction, suppose that x^* is not a Nash equilibrium of G. By generalized better-reply security, there exist $i \in N$, $U \in N(x^*_{-i})$, a well-behaved correspondence $\varphi_i : U \rightrightarrows X_i$ and $\alpha_i > u_i^*$ such that $u_i(z) \geq \alpha_i$ for all $z \in \text{graph}(\varphi_i)$. Let $x_i \in \varphi_i(x^*_{-i})$. Then, $\varphi_i \in W_U(x_i, x^*_{-i})$ and so $\underline{v}_i(x^*_{-i}) \geq \underline{u}_i(x_i, x^*_{-i}) \geq \inf_{z \in \text{graph}(\varphi_i)} u_i(z) \geq \alpha_i > u_i^*$. This is a contradiction. Hence, x^* is a Nash equilibrium of G. ∎

Lemma 3.4 establishes some properties of the game $\underline{G}$. The game $\underline{G}$ is, like G, compact and quasiconcave, but, unlike G, it is generalized payoff secure. Formally, a game $G = (X_i, u_i)_{i \in N}$ is *generalized payoff secure* if u_i is generalized payoff secure for all $i \in N$; we say that u_i, $i \in N$, is *generalized payoff secure* if for all $\varepsilon > 0$ and $x \in X$ there exists an open neighborhood $V_{x_{-i}}$ of x_{-i} and a well-behaved correspondence $\varphi_i : V_{x_{-i}} \rightrightarrows X_i$ such that $u_i(x') \geq u_i(x) - \varepsilon$ for all $x' \in \text{graph}(\varphi_i)$.

As a consequence of the generalized payoff security of $\underline{G}$, we also obtain that each player's value function in $\underline{G}$ is lower semicontinuous.

Furthermore, Lemma 3.4 also shows that $\underline{u}_i$ is below u_i for all players $i \in N$. Thus, $\underline{u}_i$ is a generalized better-reply secure approximation of u_i from below.

Lemma 3.4   *Let $G = (X_i, u_i)_{i \in N} \in \mathbb{G}_q$. Then, for all $i \in N$,*

1. $\underline{u}_i$ *is bounded,*
2. $\underline{u}_i(\cdot, x_{-i})$ *is quasi-concave for all $x_{-i} \in X_{-i}$,*
3. $\underline{u}_i$ *is generalized payoff secure,*
4. $\underline{v}_i$ *is lower semicontinuous, and*
5. $\underline{u}_i \leq u_i.$

Proof.   Let $i \in N$. Since G is compact and, in particular, u_i is bounded, it follows that $\underline{u}_i$ is also bounded.

We turn to part 2. Let $\alpha \in \mathbb{R}$, $x_i, x_i' \in \{y_i \in X_i : \underline{u}_i(y_i, x_{-i}) > \alpha\}$ and $\lambda \in (0, 1)$. Then, there exist $U, U' \in N(x_{-i})$, $\varphi_i \in W_U(x_i, x_{-i})$ and $\varphi_i' \in W_{U'}$ (x_i', x_{-i}) such that $\inf_{z \in \text{graph}(\varphi_i)} u_i(z) > \alpha$ and $\inf_{z \in \text{graph}(\varphi_i')} u_i(z) > \alpha$.

Define $\bar{x}_i = \lambda x_i + (1 - \lambda)x_i'$, $\bar{U} = U \cap U'$ and $\bar{\varphi}_i = \lambda \varphi_i + (1 - \lambda)\varphi_i'$ in $\bar{U}$. It follows that $\bar{U}$ is an open neighborhood of x_{-i} and $\bar{x}_i \in \bar{\varphi}_i(x_{-i})$. Furthermore, $\bar{\varphi}_i$ is well-behaved by Theorem A.8. Hence, $\bar{U} \in N(x_{-i})$ and $\bar{\varphi}_i \in W_{\bar{U}}(\bar{x}_i, x_{-i})$.

Let $\bar{z} \in \text{graph}(\bar{\varphi}_i)$. Then, there exist $z_i \in \varphi_i(\bar{z}_{-i})$ and $z_i' \in \varphi_i'(\bar{z}_{-i})$ such that $\bar{z}_i = \lambda z_i + (1 - \lambda)z_i'$. The quasiconcavity of $u_i(\cdot, \bar{z}_{-i})$ implies that

$$u_i(\bar{z}) \geq \min\{u_i(z_i, \bar{z}_{-i}), u_i(z_i', \bar{z}_{-i})\}$$

$$\geq \min\left\{ \inf_{y \in \text{graph}(\varphi_i)} u_i(y), \inf_{y \in \text{graph}(\varphi_i')} u_i(y) \right\}$$

and, thus, $\inf_{z \in \text{graph}(\bar{\varphi}_i)} u_i(z) > \alpha$. Hence, $\underline{u}_i(\bar{x}_i, x_{-i}) \geq \inf_{z \in \text{graph}(\bar{\varphi}_i)} u_i$ $(z) > \alpha$, which implies that $\bar{x}_i = \lambda x_i + (1 - \lambda)x_i' \in \{y_i \in X_i : \underline{u}_i(y_i, x_{-i}) > \alpha\}$ and that $\underline{u}_i(\cdot, x_{-i})$ is quasiconcave.

We next show that $\underline{u}_i$ is generalized payoff secure. Let $i \in N$, $\varepsilon > 0$ and $x \in X$. Then, there exists $U \in N(x_{-i})$ and $\varphi_i \in W_U(x)$ such that $\inf_{z \in \text{graph}(\varphi_i)} u_i(z) > \underline{u}_i(x) - \varepsilon$. Then, for all $x' \in \text{graph}(\varphi_i)$, we have that $x_{-i}' \in U$ and $x_i' \in \varphi_i(x_{-i}')$, that is, $U \in N(x_{-i}')$ and $\varphi_i \in W_U(x')$. Thus, $\underline{u}_i(x') \geq \inf_{z \in \text{graph}(\varphi_i)} u_i(z) > \underline{u}_i(x) - \varepsilon$ for all $x' \in \text{graph}(\varphi_i)$.

We next prove that $\underline{v}_i$ is lower semicontinuous. Since $\underline{u}$ is bounded, it follows that $\underline{v}_i$ is real-valued for all $i \in N$. Let $i \in N$, $x_{-i} \in X_{-i}$ and $\varepsilon > 0$ be given. Let $0 < \eta < \varepsilon$ and let $x_i \in X_i$ be such that $\underline{u}_i(x_i, x_{-i}) - \eta > \underline{v}_i(x_{-i}) - \varepsilon$. Since $\underline{u}_i$ is generalized payoff secure, there exists an open neighborhood $V_{x_{-i}}$ of x_{-i} and a well-behaved correspondence $\varphi_i : V_{x_{-i}} \rightrightarrows X_i$ such that $\underline{u}_i(x') \geq \underline{u}_i(x) - \eta$ for all $x' \in \operatorname{graph}(\varphi_i)$. Then, for all $x'_{-i} \in V_{x_{-i}}$, letting $x'_i \in \varphi_i(x'_{-i})$, we obtain that $\underline{v}_i(x'_{-i}) \geq \underline{u}_i(x') \geq \underline{u}_i(x) - \eta > \underline{v}_i(x_{-i}) - \varepsilon$. Hence, $\underline{v}_i$ is lower semicontinuous.

Finally, we show that $u_i(x) \geq \underline{u}_i(x)$ for all $i \in N$ and $x \in X$. In fact, for all $U \in N(x_{-i})$ and $\varphi_i \in W_U(x_i, x_{-i})$, we have that $x \in \operatorname{graph}(\varphi_i)$. Thus, $\inf_{z \in \operatorname{graph}(\varphi_i)} u_i(z) \leq u_i(x)$ and so $\underline{u}_i(x) \leq u_i(x)$. ∎

The next lemma shows that $\underline{G}$ has an f-equilibrium provided that f is continuous, strictly below $\underline{v}$ and, like $\underline{v}_i$, f_i depends only on x_{-i} for all $i \in N$. Note that Lemma 3.5 can be understood as stating that every compact, quasiconcave and generalized payoff secure game has an f-equilibrium for all continuous f that approximate players' value functions strictly from below.

Lemma 3.5   *Let $G = (X_i, u_i)_{i \in N} \in \mathbb{G}_q$. Then $\underline{G} = (X_i, \underline{u}_i)_{i \in N}$ has an f-equilibrium for all continuous $f \in F(\underline{G})$ satisfying $f_i(x) = f_i(x'_i, x_{-i})$ and $f_i(x) < \underline{v}_i(x_{-i})$ for all $i \in N$, $x'_i \in X_i$ and $x \in X$.*

Proof.   Let $f \in F(\underline{G})$ be such that $f_i(x) = f_i(x'_i, x_{-i})$ and $f_i(x_{-i}) < \underline{v}_i(x_{-i})$ for all $i \in N$, $x'_i \in X_i$ and $x \in X$. In particular, we may write $f_i(x_{-i})$ instead of $f_i(x)$. Define $\Psi : X \rightrightarrows X$ by $\Psi(x) = \{y \in X : \underline{u}_i(y_i, x_{-i}) > f_i(x_{-i})$ for all $i \in N\}$. Note that Ψ is nonempty-valued since $f_i < \underline{v}_i$ for all $i \in N$ and is convex-valued since $\underline{u}_i(\cdot, x_{-i})$ is quasiconcave for all $i \in N$ and $x_{-i} \in X_{-i}$ (by Lemma 3.4).

Next, we show that for all $x \in X$, there exist an open neighborhood V_x of x and a well-behaved correspondence $\varphi_x : V_x \rightrightarrows X$ such that $\varphi_x(x') \subseteq \Psi(x')$ for all $x' \in V_x$.

In order to establish the above claim, let $x \in X$ and consider $y \in \Psi(x)$. Fix $i \in N$. Then $\underline{u}_i(y_i, x_{-i}) > f_i(x_{-i}) + 2\eta$ for some $\eta > 0$ sufficiently small. Since $\underline{G}$ is generalized payoff secure and f is continuous, it follows that there exist an open neighborhood $V_{x_{-i}}$ of x_{-i} and a well-behaved correspondence $\varphi_i : V_{x_{-i}} \rightrightarrows X_i$ such that $\underline{u}_i(x') \geq \underline{u}_i(y_i, x_{-i}) - \eta$ for all $x' \in \operatorname{graph}(\varphi_i)$ and $f_i(x_{-i}) > f_i(x'_{-i}) - \eta$ for all $x'_{-i} \in V_{x_{-i}}$. Define $V_i = X_i \times V_{x_{-i}}$; furthermore, define $V_x = \cap_{i \in N} V_i$ and $\varphi_x : V \rightrightarrows X$ by $\varphi_x(x') = \prod_{i \in N} \varphi_i(x'_{-i})$ for all $x' \in V$. Let $x' \in V_x$ and $y' \in \varphi_x(x')$. Then, for all $i \in N$, it follows that $x'_{-i} \in V_{x_{-i}}$ and $y'_i \in \varphi_i(x'_{-i})$. Then $\underline{u}_i(y'_i, x'_{-i}) \geq \underline{u}_i(y_i, x_{-i}) - \eta > f_i(x_{-i}) + \eta > f_i(x'_{-i})$ for all $i \in N$. Hence, $y' \in \Psi(x')$ and so $\varphi_x(x') \subseteq \Psi(x')$.

In the light of the above claim, we obtain a family $\{V_x\}_{x \in X}$ where V_x is an open neighborhood of x, and a family $\{\varphi_x\}_{x \in X}$ where $\varphi_x : V_x \rightrightarrows X$ is a well-behaved correspondence satisfying $\varphi_x(x') \subseteq \Psi(x')$ for all $x' \in V_x$. Since X is compact, there exist a finite open cover $\{V_{x_j}\}_{j=1}^m$ and, by Theorem A.3, a partition of unity $\{\beta_j\}_{j=1}^m$ subordinate to $\{V_{x_j}\}_{j=1}^m$. Define $\phi : X \rightrightarrows X$ by $\phi(x) = \sum_{j=1}^m \beta_j(x)\varphi_{x_j}(x)$. Then, ϕ is a well-behaved correspondence by Theorem A.8 and, therefore, ϕ has a fixed point x^* by Theorem A.14.

Let $j \in \{1, \ldots, m\}$. We have that $\varphi_{x_j}(x^*) \subseteq \Psi(x^*)$ if $x^* \in V_{x_j}$, and $\beta_j(x^*) = 0$ if $x^* \notin V_{x_j}$. Since Ψ is convex-valued, then $x^* \in \phi(x^*)$ $= \sum_{j:\beta_j(x^*)>0} \beta_j(x^*)\varphi_{x_j}(x^*) \subseteq \Psi(x^*)$. Hence, for all $i \in N$, $\underline{u}_i(x^*) > f_i(x^*_{-i})$ and so x^* is an f-equilibrium of $\underline{G}$. ∎

Lemma 3.5 shows why it is appealing, from the viewpoint of existence of equilibrium, to have generalized payoff secure games: all such games have f-equilibria provided that f is as in its statement. A second reason why generalized payoff security is appealing, which in fact reinforces the first one, is that $\underline{v}_i$ is lower semicontinuous for all $i \in N$. In fact, the lower semicontinuity of $\underline{v}_i$ for all $i \in N$ implies, together with Theorem A.1, that there exists a sequence $\{v_i^k\}_{k=1}^\infty$ of continuous real-valued functions on X_{-i} such that $v_i^k(x_{-i}) \leq \underline{v}_i(x_{-i})$ and $\liminf_k v_i^k(x^k_{-i}) \geq \underline{v}_i(x_{-i})$ for all $k \in \mathbb{N}$, $i \in N$, $x_{-i} \in X_{-i}$ and all sequences $\{x^k_{-i}\}_{k=1}^\infty$ converging to x_{-i}. This form of approximation of players' value functions together with generalized better-reply security is enough for every limit point of every sequence of approximate equilibria of $\underline{G}$ to be a Nash equilibrium of G.

Lemma 3.6 Let $G = (X_i, u_i)_{i \in N} \in \mathbb{G}_q$ be generalized better-reply secure. If $x^* \in X$, $\{x_k\}_{k=1}^\infty \subseteq X$ and $\{f_k\}_{k=1}^\infty \subseteq F(\underline{G})$ are such that $x^* = \lim_k x_k$, $\liminf_k f_i^k(x_k) \geq \underline{v}_i(x^*_{-i})$ for all $i \in N$ and x_k is a f_k-equilibrium of $\underline{G} = (X_i, \underline{u}_i)_{i \in N}$ for all $k \in \mathbb{N}$, then x^* is a Nash equilibrium of G.

Proof. Since u is bounded, taking a subsequence if necessary, we may assume that $\{u(x_k)\}_{k=1}^\infty$ converges. Let $u^* = \lim_k u(x_k)$ and note that $(x^*, u^*) \in \Gamma$.

Let $i \in N$. Since x_k is a f_k-equilibrium of $\underline{G}$ for all $k \in \mathbb{N}$, then $u_i(x_k) \geq \underline{u}_i(x_k) \geq f_i^k(x_k)$ for all $k \in \mathbb{N}$, and so $\liminf_k f_i^k(x_k) \geq \underline{v}_i(x^*_{-i})$ implies that $u_i^* \geq \underline{v}_i(x^*_{-i})$. Since $u_i^* \geq \underline{v}_i(x^*_{-i})$ for all $i \in N$ and G is generalized better-reply secure, it follows by Theorem 3.3 that x^* is a Nash equilibrium of G. ∎

Note that Lemma 3.6 generalizes Theorem 2.5 by weakening the continuity assumption on G to generalized better-reply security.

We finally turn to the proof of Theorem 3.2, which is obtained easily from Lemmas 3.4–3.6.

Proof of Theorem 3.2.   For all $k \in \mathbb{N}$ and $i \in N$, let $f_i^k : X_{-i} \to \mathbb{R}$ be defined by $f_i^k(x_{-i}) = v_i^k(x_{-i}) - 1/k$ for all $x_{-i} \in X_{-i}$, where $\{v_i^k\}_{k=1}^\infty$ is a sequence of continuous real-valued functions on X_{-i} such that $v_i^k(x_{-i}) \leq \underline{v}_i(x_{-i})$ and $\liminf_k v_i^k(x_{-i}^k) \geq \underline{v}_i(x_{-i})$ for all $k \in \mathbb{N}$, $i \in N$, $x_{-i} \in X_{-i}$ and all sequences $\{x_{-i}^k\}_{k=1}^\infty$ converging to x_{-i} (as remarked above, the existence of this sequence follows from Lemma 3.4 and Theorem A.1). Since f_k is continuous and $f_i^k < \underline{v}_i$ for all $i \in N$, Lemma 3.5 implies that $\underline{G}$ has a f_k-equilibrium, x_k, for all $k \in \mathbb{N}$.

Since X is compact, we may assume that $\{x_k\}_{k=1}^\infty$ converges. Letting $x^* = \lim_k x_k$, we have that $\liminf_k f_i^k(x_{-i}^k) \geq \underline{v}_i(x_{-i}^*)$ for all $i \in N$ and Lemma 3.6 implies that x^* is a Nash equilibrium of G. ∎

3.2   Examples

The following two examples illustrate the notion of generalized better-reply security and the existence result this notion allows.

The first example can be described as an imitation game. Suppose that G is such that there are two players, $N = \{1, 2\}$, who have the same (compact, convex, metric) strategy space $X_1 = X_2 = A$. Player 1's payoff function $u_1 : X \to \mathbb{R}$ is continuous and quasiconcave in x_1 and player 2's payoff function is

$$u_2(x) = \begin{cases} 1 & \text{if } x_1 = x_2, \\ 0 & \text{otherwise.} \end{cases}$$

We can interpret player 2's payoff function as representing a situation where player 2 wants to imitate player 1.

We start by showing that, in this example, $\underline{u}_i = u_i$ for all $i \in N$. Since, by Lemma 3.4, we have that $\underline{u}_i \leq u_i$, it suffices to show that for all $x \in X$ and $\varepsilon > 0$, $\underline{u}_i(x) \geq u_i(x) - \varepsilon$.

For player 1, given $x \in X$ and $\varepsilon > 0$, the continuity of u_1 implies the existence of $U \in N(x_2)$ such that $u_1(x_1, x_2') > u_1(x) - \varepsilon$ for all $x_2' \in U$. Hence, letting φ_1 denote the constant correspondence equal to $\{x_1\}$ on U, it follows that $\varphi_1 \in W_U(x)$ and that $\underline{u}_1(x) \geq \inf_{z \in \text{graph}(\varphi)} u_1(z) \geq u_1(x) - \varepsilon$.

Regarding player 2, given $x \in X$ and $\varepsilon > 0$, let $U = X_1$ and $\varphi_2(x_1') = \{x_1'\}$ for all $x_1' \in U$. Then, $U \in N(x_1)$, $\varphi_2 \in W_U(x)$ and $\underline{u}_2(x) \geq \inf_{z \in \text{graph}(\varphi)} u_2(z) = 1 > u_2(x) - \varepsilon$.

We next show that G is generalized better-reply secure. Let $(x^*, u^*) \in \Gamma$ be such that $u_i^* \geq \underline{v}_i(x_{-i}^*) = v_i(x_{-i}^*)$ for all $i \in N$. Since u_1 is continuous, it follows that $u_1^* = u_1(x^*)$ and, hence, $u_1(x^*) \geq v_1(x_2^*)$. Furthermore, since $v_2(x_1^*) = 1$, then $u_2^* = 1$. Letting $\{x_k\}_{k=1}^\infty \subseteq X$ be such that $\lim_k (x_k, u(x_k)) = (x^*, u^*)$, we have that $u_2(x_k) = 1$ for all k sufficiently large. Thus, $x_2^k = x_1^k$ for all k sufficiently large and, hence, $x_2^* = x_1^*$. Thus, $u_2(x^*) = 1 = v_2(x_1^*)$. Since we have $u_i(x^*) \geq v_i(x_{-i}^*)$ for all $i \in N$, it follows that x^* is a Nash equilibrium of G. Hence, G is generalized better-reply secure. Furthermore, it follows from Theorem 3.2 that G has a Nash equilibrium.

The second example considers the pure exchange general equilibrium model. There are n consumers, each of whom can consume m commodities. Each consumer $i \in \{1, \ldots, n\}$ is characterized by a continuous, strictly increasing, quasiconcave utility function $u_i : \mathbb{R}_+^m \to \mathbb{R}_+$ and a strictly positive endowment vector e_i. A *competitive equilibrium* is $(p^*, x^*) \in \mathbb{R}_+^m \times \mathbb{R}_+^{mn}$ such that

(a) x_i^* solves $\max_{x_i \in \mathbb{R}_+^m} u_i(x_i)$ subject to $p^* \cdot x_i \leq p^* \cdot e_i$, and
(b) $\sum_{i=1}^n x_i^* = \sum_{i=1}^n e_i$.

The existence of a competitive equilibrium for the above pure exchange economy will be established via the existence of a Nash equilibrium of the following game played by the consumers and an auctioneer (player 0). Let $G = (X_i, w_i)_{i \in N}$ be defined by $N = \{0, \ldots, n\}$,

$$X_0 = \left\{ p \in \mathbb{R}_+^m : \sum_{j=1}^m p_j = 1 \right\},$$

$$X_i = \left\{ x_i \in \mathbb{R}_+^m : x_i^j \leq 1 + \sum_{l=1}^n e_l^j \text{ for all } j = 1, \ldots, m \right\}$$

$$\text{for all } i \in \{1, \ldots, n\},$$

$$w_0(x_0, x_1, \ldots, x_n) = x_0 \cdot \sum_{i=1}^n (x_i - e_i)$$

and, for all $i \in \{1, \ldots, n\}$,

$$w_i(x_0, x_1, \ldots, x_n) = \begin{cases} u_i(x_i) & \text{if } x_0 \cdot x_i \leq x_0 \cdot e_i, \\ -1 & \text{otherwise.} \end{cases}$$

It is easy to check that G is quasiconcave, compact and metric. We next show that G is generalized better-reply secure. Let $(x^*, u^*) \in \mathrm{cl}(\mathrm{graph}(w))$ be such that x^* is not a Nash equilibrium of G. Hence, there exists $i \in N$ and $x_i \in X_i$ such that $w_i(x_i, x^*_{-i}) > w_i(x^*)$ and, furthermore, there exists $\varepsilon > 0$ such that $w_i(x_i, x^*_{-i}) > w_i(x^*) + \varepsilon$. We consider several possible cases.

Suppose that $i = 0$. Since w_0 is continuous, then $u_0^* = w_0(x^*)$ and there exists $U \in N(x^*_{-0})$ such that $w_0(x_0, x_{-0}) > w_0(x^*) + \varepsilon = u_0^* + \varepsilon$ for all $x_{-0} \in U$. Hence, in this case, G is generalized better-reply secure.

Suppose next that $i \in \{1, \ldots, n\}$. If $u_i^* = -1$, then simply let $\varphi_i : X_{-i} \rightrightarrows X_i$ be defined by $\varphi_i(x_{-i}) = \{x_i \in X_i : x_0 \cdot x_i \le x_0 \cdot e_i\}$. Then, φ_i is well-behaved and $w_i(z) \ge 0 > -1 = u_i^*$ for all $z \in \mathrm{graph}(\varphi_i)$.

Thus, we may assume that $u_i^* \ge 0$. This implies that there is a sequence $\{x_k\}_{k=1}^{\infty}$ such that $\lim_k x_k = x^*$ and $x_0^k \cdot x_i^k \le x_0^k \cdot e_i$ for all k. Hence, $u_i^* = \lim_k u_i(x_i^k) = u_i(x_i^*)$ and $x_0^* \cdot x_i^* \le x_0^* \cdot e_i$. The latter condition implies that $w_i(x^*) = u_i(x_i^*)$ and, since $w_i(x_i, x^*_{-i}) > w_i(x^*) + \varepsilon$, we also have that $w_i(x_i, x^*_{-i}) = u_i(x_i)$. In particular, $u_i(x_i) > u_i(x_i^*) + \varepsilon$.

Let $\lambda \in (0, 1)$ be such that $u_i(\lambda x_i) > u_i(x_i^*) + \varepsilon$. Since $x_0^* \cdot x_i \le x_0^* \cdot e_i$, then $x_0^* \cdot \lambda x_i < x_0^* \cdot e_i$ and, therefore, there exists $V \in N(x_0^*)$ such that $x_0 \cdot \lambda x_i < x_0 \cdot e_i$ for all $x_0 \in V$. Letting $U = V \times \prod_{l \neq 0,i} X_l$ and $\varphi_i(x_{-i}) = \{\lambda x_i\} \subseteq X_i$ for all $x_{-i} \in U$, we have that $w_i(z) = u_i(\lambda x_i) > u_i(x_i^*) + \varepsilon = u_i^* + \varepsilon$ for all $z \in \mathrm{graph}(\varphi_i)$. Thus, G is generalized better-reply secure.

Having established that $G \in \mathbb{G}_q$ is generalized better-reply secure, it follows by Theorem 3.2 that G has a Nash equilibrium $(x_0^*, \ldots, x_n^*)$. For convenience, let $p^* = x_0^*$ and $x^* = (x_1^*, \ldots, x_n^*)$. We next show that (p^*, x^*) is a competitive equilibrium.

Consider $i \in \{1, \ldots, n\}$. Since $0 \in X_i$ is such that $w_i(p^*, 0, x^*_{-i}) = u_i(0) \ge 0$, it follows that $p^* \cdot x_i^* \le p^* \cdot e_i$; otherwise, $w_i(p^*, x^*) = -1 < w_i(p^*, 0, x^*_{-i})$. Moreover, for all $x_i \in X_i$ such that $p^* \cdot x_i \le p^* \cdot e_i$, we have that $u_i(x_i^*) = w_i(p^*, x^*) \ge w_i(p^*, x_i, x^*_{-i}) = u_i(x_i)$. Hence, (a) holds.

Furthermore, since u_i is strictly increasing, then $p^* > 0$ (otherwise there is no solution to the maximization problem in (a)) and $p^* \cdot x_i^* = p^* \cdot e_i$ (otherwise x_i^* is not a solution to the maximization problem in (a)). Thus, $w_0(p^*, x^*) = \sum_{i=1}^{n}(p^* \cdot x_i^* - p^* \cdot e_i) = 0$. Since we also have $0 = w_0(p^*, x^*) = p^* \cdot \sum_{i=1}^{n}(x_i^* - e_i)$, then $p^* > 0$ implies that that $\sum_{i=1}^{n}(x_i^* - e_i) = 0$. Hence, (b) also holds.

We conclude this section by noting that the above examples suggest conditions that are easy to verify and sufficient for generalized better-reply security. Some of these conditions will, indeed, be presented in Section 3.5.

3.3 Two Characterizations of Generalized Better-Reply Security

We have shown in Section 3.1 that, given a normal-form game $G = (X_i, u_i)_{i \in N}$, the function $\underline{u}$ is generalized payoff secure, approximates u from below and is tied to u through the defining property of generalized better-reply security (this is the sense in which $\underline{u}$ approximates u). Other functions can conceivably approximate u in a similar way and yield a stronger existence result. Motivated by this observation, we will formalize the notion of weak better-reply secure relative to a generalized payoff secure function $\tilde{u}$ below u in a way that a game is weak better-reply secure relative to $\underline{u}$ if and only if the game is generalized better-reply secure. We then show that our conjecture is false by showing that a game is weakly better-reply secure relative to some $\tilde{u}$ if and only if it is weakly better-reply secure relative to $\underline{u}$. In other words, $\underline{u}$ is the best approximation of u having the above properties.

Weak better-reply security is defined as follows. Let $G = (X_i, u_i)_{i \in N}$ be a normal-form game and $\tilde{u}$ be a bounded $\mathbb{R}^n$-valued function on X; furthermore, let $\tilde{v}_i(x_{-i}) = \sup_{x_i \in X_i} \tilde{u}_i(x_i, x_{-i})$ for all $i \in N$ and $x_{-i} \in X_{-i}$. We say that G is *weakly better-reply secure relative to $\tilde{u}$* if

(a) $\tilde{u}_i \leq u_i$ for all $i \in N$,
(b) $\tilde{u}_i$ is generalized payoff secure for all $i \in N$,
(c) $\tilde{u}_i(\cdot, x_{-i})$ is quasiconcave for all $i \in N$ and $x_{-i} \in X_{-i}$, and
(d) x^* is a Nash equilibrium of G for all $(x^*, u^*) \in \Gamma$ such that $u_i^* \geq \tilde{v}_i(x_{-i}^*)$.

Moreover, we say that G is *weakly better-reply secure* if there exists a bounded $\mathbb{R}^n$-valued function $\tilde{u}$ on X such that G is weakly better-reply secure relative to $\tilde{u}$.

Theorem 3.7 shows that weak better-reply security relative to $\underline{u}$ is the minimal form of weak better-reply security. Furthermore, it shows that weak better-reply security is equivalent to generalized better-reply security.

Theorem 3.7   *Let $G = (X_i, u_i)_{i \in N} \in \mathbb{G}_q$. Then, the following conditions are equivalent:*

1. *G is weakly better-reply secure.*
2. *G is weakly better-reply secure relative to $\underline{u}$.*
3. *G is generalized better-reply secure.*

Theorem 3.7 is obtained from Lemma 3.8 below. We already know from Lemma 3.4 that $\underline{u}$ approximates u from below and is generalized payoff secure. Lemma 3.8 shows that $\underline{u}_i$ is the pointwise supremum of the set of functions $\tilde{u}$ that approximate u from below and are generalized payoff secure. In other words, if $\tilde{u}$ approximates u from below and is generalized payoff secure, then $\tilde{u}_i$ is (everywhere) below $\underline{u}_i$. For this reason, we say that $\underline{u}_i$ is the *generalized payoff secure envelope* of u_i.

The following notation is used in the statement of Lemma 3.8. Let $G = (X_i, u_i)_{i \in N}$ be a game and define $L(G)$ as the set of all bounded functions $\tilde{u} : X \to \mathbb{R}^n$ such that $\tilde{u}_i \leq u_i$ and $\tilde{u}_i$ is generalized payoff secure for all $i \in N$.

Lemma 3.8   *Let $G = (X_i, u_i)_{i \in N} \in \mathbb{G}_q$. Then $\underline{u} \in L(G)$ and $\underline{u} \geq \tilde{u}$ for all $\tilde{u} \in L(G)$.*

Proof.   First note that $\underline{u} \in L(G)$ by Lemma 3.4. We next establish that $\underline{u}_i(x) \geq \tilde{u}_i(x)$ for all $\tilde{u} \in L(G)$, $i \in N$ and $x \in X$. Let $\tilde{u} \in L(G)$, $i \in N$, $x \in X$ and $\varepsilon > 0$ be given. Since $\tilde{u} \in L(G)$, then there exists $U \in N(x_{-i})$ and a well-behaved correspondence $\varphi_i : U \rightrightarrows X_i$ such that $\tilde{u}_i(z) \geq \tilde{u}_i(x) - \varepsilon$ for all $z \in \text{graph}(\varphi_i)$. Since $u_i \geq \tilde{u}_i$, then $u_i(z) \geq \tilde{u}_i(x) - \varepsilon$ for all $z \in \text{graph}(\varphi_i)$.

Define $\psi_i : U \rightrightarrows X_i$ by

$$\psi_i(x'_{-i}) = \begin{cases} \text{co}(\{x_i\} \cup \varphi_i(x_{-i})) & \text{if } x'_{-i} = x_{-i}, \\ \varphi_i(x_{-i}) & \text{otherwise.} \end{cases}$$

Then ψ_i is well-behaved and $x \in \text{graph}(\psi_i)$, i.e., $\psi_i \in W_U(x)$. Furthermore, for all $z \in \text{graph}(\psi_i)$, either (a) $z = x$, or (b) $z \in \text{graph}(\varphi_i)$ or (c) $z = \lambda x + (1 - \lambda)(z'_i, x_{-i})$ for some $\lambda \in (0, 1)$ and $z'_i \in \varphi_i(x_{-i})$. Thus, if $z \in \text{graph}(\psi_i)$, then $u_i(z) = u_i(x) \geq \tilde{u}_i(x) > \tilde{u}_i(x) - \varepsilon$ in case (a), $u_i(z) \geq \tilde{u}_i(x) - \varepsilon$ in case (b) and, due to the quasi-concavity of $u_i(\cdot, x_{-i})$, $u_i(z) \geq \min\{u_i(x), u_i(z'_i, x_{-i})\} \geq \tilde{u}_i(x) - \varepsilon$ in case (c). Hence, $\underline{u}_i(x) \geq \inf_{z \in \text{graph}(\psi_i)} u_i(z) \geq \tilde{u}_i(x) - \varepsilon$. Since $\varepsilon > 0$ is arbitrary, it follows that $\underline{u}_i(x) \geq \tilde{u}_i(x)$.   ∎

Theorem 3.7 follows easily from Lemma 3.8.

Proof of Theorem 3.7.   Theorem 3.3 and Lemma 3.4 show that conditions 2 and 3 are equivalent. Furthermore, it is clear that condition 2 implies condition 1. Hence, it suffices to show that condition 3 is implied by condition 1.

Suppose that G is weakly better-reply secure relative to $\tilde{u}$ and consider $(x^*, u^*) \in \Gamma$ such that $u_i^* \geq \underline{v}_i(x_{-i}^*)$ for all $i \in N$. Since G is weakly better-reply secure relative to $\tilde{u}$, then $\tilde{u} \in L(G)$ and, it follows by Lemma 3.8 that $\underline{v}_i(x_{-i}^*) \geq \tilde{v}_i(x_{-i}^*)$ for all $i \in N$. Hence, $(x^*, u^*) \in \Gamma$ is such that $u_i^* \geq \tilde{v}_i(x_{-i}^*)$ for all $i \in N$ and, since G is weakly better-reply secure relative to $\tilde{u}$, x^* is a Nash equilibrium of G. Thus, G is generalized better-reply secure. ∎

We next present a condition, reducible security, that is related to weak better-reply security. This condition shares with weak better-reply security the property that players' payoff functions are replaced with "approximating" generalized better-reply secure function. However, reducible security differs from generalized better-reply security because, first, the approximation is not required to be from below and, second, the way the approximation is related with the original payoff function is different in the two conditions.

Let $G = (X_i, u_i)_{i \in N}$ be a normal-form game and w be a bounded $\mathbb{R}^n$-valued function on X; furthermore, let $v_{w_i}(x_{-i}) = \sup_{x_i \in X_i} w_i(x_i, x_{-i})$ for all $i \in N$ and $x_{-i} \in X_{-i}$. We say that G is *reducible secure relative to w* if

(a) w_i is generalized payoff secure for all $i \in N$,
(b) $w_i(\cdot, x_{-i})$ is quasiconcave for all $i \in N$ and $x_{-i} \in X_{-i}$, and
(c) if whenever $x \in X$ is not a Nash equilibrium of G, there exists $\varepsilon > 0$ and a neighborhood U of x such that, for all $x' \in U$, there is a player $i \in N$ for whom $v_{w_i}(x_{-i}) \geq w_i(x') + \varepsilon$.

Moreover, we say that G is *reducible secure* if there exists a bounded $\mathbb{R}^n$-valued function w on X such that G is reducible secure relative to w.

The notion of reducible security is not useful to address existence of Nash equilibrium because it happens that the games that are reducible secure are precisely those that have Nash equilibria. In other words, a game G is reducible secure if and only if $E(G)$ is nonempty (see Carmona (2011a)). This means that, in order for reducible security to be useful, one must restrict the set of the functions w to which games can be reducible secure relative to.

Theorem 3.9 consider reducible security with respect to a particular function, namely, the generalized payoff secure envelope $\underline{u}$ of players' payoff functions u. It shows that reducible security relative to $\underline{u}$ is precisely equivalent to generalized better-reply security.

Theorem 3.9   *Let $G = (X_i, u_i)_{i \in N} \in \mathbb{G}_q$. Then, G is generalized better-reply secure if and only if G is reducible secure relative to $\underline{u}$.*

Proof.   (Sufficiency) Suppose that G is not reducible secure relative to $\underline{u}$. We will show that G is not generalized better-reply secure.

Since G is not reducible secure relative to $\underline{u}$, there exists $x^* \notin E(G)$ such that for all $\varepsilon > 0$ and $U \in N(x^*)$ there exists $x' \in U$ such that $\underline{v}_i(x^*_{-i}) < \underline{u}_i(x') + \varepsilon$ for all $i \in N$. Letting, for all $k \in \mathbb{N}$, $\varepsilon = 1/k$ and U be the open ball of radius $1/k$ around x^*, we obtain a sequence $\{x_k\}_{k=1}^{\infty}$ such that $\lim_k x_k = x^*$ and $\underline{v}_i(x^*_{-i}) < \underline{u}_i(x_k) + 1/k$ for all $i \in N$.

Since u is bounded, taking a subsequence if necessary, we may assume that $\{u(x_k)\}_k$ converges. Let $u^* = \lim_k u(x_k)$ and note that $(x^*, u^*) \in \mathrm{cl}(\mathrm{graph}(u))$. Furthermore, since $\underline{u} \leq u$, then $\underline{v}_i(x^*_{-i}) < u_i(x_k) + 1/k$ for all $k \in \mathbb{N}$ and $i \in N$. Hence, $\underline{v}_i(x^*_{-i}) \leq u^*_i$ for all $i \in N$. Since $x^* \notin E(G)$, this implies that G is not generalized better-reply secure.

(Necessity) Suppose that G is reducible secure relative to $\underline{u}$. To show that G is generalized better-reply secure, it suffices, by Theorem 3.3, to show that for all $(x^*, u^*) \in \mathrm{cl}(\mathrm{graph}(u))$ such that $x^* \notin E(G)$, there exists $i \in N$ such that $\underline{v}_i(x^*_{-i}) > u^*_i$.

Let $(x^*, u^*) \in \mathrm{cl}(\mathrm{graph}(u))$ be such that $x^* \notin E(G)$. Then, since G is reducible secure relative to $\underline{u}$, there exists $\varepsilon > 0$ and an open neighborhood U of x^* such that for all $x' \in U$ there exists $i \in N$ such that $\underline{v}_i(x^*_{-i}) \geq \underline{u}_i(x') + \varepsilon$.

Let $\{x_k\}_{k=1}^{\infty}$ be a sequence such that $\lim_k (x_k, u(x_k)) = (x^*, u^*)$ and such that $\{\underline{u}(x_k)\}_{k=1}^{\infty}$ converges. Let $\beta = \lim_k \underline{u}(x_k)$. Since we eventually have $x_k \in U$ and there are only finitely many players, then there exists $i \in N$ such that

$$\underline{v}_i(x^*_{-i}) \geq \beta_i + \varepsilon. \tag{3.1}$$

Let $0 < \eta < \varepsilon$. Since $\eta > 0$, there exists $y_i \in X_i$ such that $\underline{u}_i(y_i, x^*_{-i}) > \underline{v}_i(x^*_{-i}) - \eta$. Hence, there exist an open neighborhood V of x^*_{-i} and a well-behaved correspondence $\varphi_i : V \rightrightarrows X_i$ such that $u_i(x') \geq \underline{v}_i(x^*_{-i}) - \eta$ for all $x' \in \mathrm{graph}(\varphi_i)$. Fix $k \in \mathbb{N}$ with the property that $x^k_{-i} \in V$ and define $\psi_i : V \rightrightarrows X_i$ by

$$\psi_i(x_{-i}) = \begin{cases} \mathrm{co}(\{x^k_i\} \cup \varphi_i(x^k_{-i})) & \text{if } x_{-i} = x^k_{-i}, \\ \varphi_i(x_{-i}) & \text{otherwise.} \end{cases}$$

Then ψ_i is well-behaved and $x_k \in \mathrm{graph}(\psi_i)$. Furthermore, for all $z \in \mathrm{graph}(\psi_i)$, either (a) $z = x_k$, or (b) $z \in \mathrm{graph}(\varphi_i)$ or (c) $z = \lambda x_k + (1 - \lambda)$

(z'_i, x^k_{-i}) for some $\lambda \in (0,1)$ and $z'_i \in \varphi_i(x^k_{-i})$. Thus, if $z \in \text{graph}(\psi_i)$, then $u_i(z) = u_i(x_k)$ in case (a), $u_i(z) \geq \underline{v}_i(x^*_{-i}) - \eta$ in case (b) and, due to the quasiconcavity of $u_i(\cdot, x^k_{-i})$, $u_i(z) \geq \min\{u_i(x_k), \underline{v}_i(x^*_{-i}) - \eta\}$ in case (c). Hence, $\underline{u}_i(x_k) \geq \inf_{z \in \text{graph}(\psi_i)} u_i(z) \geq \min\{u_i(x_k), \underline{v}_i(x^*_{-i}) - \eta\}$. Since $k \in \mathbb{N}$ is arbitrary, this inequality holds for all $k \in \mathbb{N}$ such that $x^k_{-i} \in V$ and, in particular, for all k sufficiently large.

Suppose, in order to reach a contradiction, that $u^*_i \geq \underline{v}_i(x^*_{-i})$. Then, for all k sufficiently large, $\min\{u_i(x_k), \underline{v}_i(x^*_{-i}) - \eta\} = \underline{v}_i(x^*_{-i}) - \eta$ and, therefore, $\underline{u}_i(x_k) \geq \underline{v}_i(x^*_{-i}) - \eta$. Thus, $\beta_i \geq \underline{v}_i(x^*_{-i}) - \eta > \underline{v}_i(x^*_{-i}) - \varepsilon$ contradicting (3.1). Hence, it follows that $\underline{v}_i(x^*_{-i}) > u^*_i$, as desired. ∎

3.4 Better-Reply Security

Generalized better-reply security, as its name indicates, extends the notion of better-reply security which we now consider. A game $G = (X_i, u_i)_{i \in N}$ is *better-reply secure* if whenever $(x^*, u^*) \in \Gamma$ and x^* is not a Nash equilibrium, there exists a player $i \in N$, a strategy $\bar{x}_i \in X_i$, an open neighborhood U of x^*_{-i} and a number $\alpha_i > u^*_i$ such that $u_i(\bar{x}_i, x'_{-i}) \geq \alpha_i$ for all $x'_{-i} \in U$.

Better-reply security can be characterized in terms of weak better-reply security relative to a particular function. This function is similar to the generalized payoff secure envelope of players' payoff function and, as Theorem 3.12 below shows, is the best approximation from below satisfying a stronger version of generalized payoff security. Its definition is as follows. Given $G = (X_i, u_i)_{i \in N} \in \mathbb{G}_q$, let $\bar{u} : X \to \mathbb{R}^n$ be defined by

$$\bar{u}_i(x) = \sup_{U \in N(x_{-i})} \inf_{z_{-i} \in U} u_i(x_i, z_{-i})$$

for all $i \in N$ and $x \in X$. Also, let $\bar{v}_i(x_{-i}) = \sup_{x_i \in X_i} \bar{u}_i(x_i, x_{-i})$ for all $i \in N$ and $x_{-i} \in X_{-i}$.

First, we provide a characterization of better-reply security which explicitly uses $\bar{u}$ through its value function.

Theorem 3.10   *Let $G = (X_i, u_i)_{i \in N} \in \mathbb{G}_q$. Then, G is better-reply secure if and only if x^* is a Nash equilibrium of G for all $(x^*, u^*) \in \Gamma$ such that $u^*_i \geq \bar{v}_i(x^*_{-i})$ for all $i \in N$.*

Proof.   (Sufficiency) Let $G \in \mathbb{G}_q$ be such that x^* is a Nash equilibrium of G for all $(x^*, u^*) \in \Gamma$ such that $u^*_i \geq \bar{v}_i(x^*_{-i})$ for all $i \in N$. Let $(x^*, u^*) \in \Gamma$ be such that x^* is not a Nash equilibrium. Thus, there is $i \in N$

such that $\bar{v}_i(x^*_{-i}) > u^*_i$. Hence, there exists $\alpha_i > u^*_i$ and $x_i \in X_i$ such that $\bar{u}_i(x_i, x^*_{-i}) > \alpha_i$. This, in turn, implies that there exist $U \in N(x^*_{-i})$ such that $u_i(x_i, x_{-i}) > \alpha_i$ for all $x_{-i} \in U$. Thus, G is better-reply secure.

(Necessity) Suppose that $G \in \mathbb{G}_q$ is better-reply secure. Consider $(x^*, u^*) \in \Gamma$ such that $u^*_i \geq \bar{v}_i(x^*_{-i})$ for all $i \in N$, and, in order to reach a contradiction, suppose that x^* is not a Nash equilibrium of G. By better-reply security, there exist $i \in N$, $x_i \in X_i$, $U \in N(x^*_{-i})$ and $\alpha_i > u^*_i$ such that $u_i(x_i, x_{-i}) \geq \alpha_i$ for all $x_{-i} \in U$. Hence, $\bar{v}_i(x^*_{-i}) \geq \bar{u}_i(x_i, x^*_{-i}) \geq \inf_{x_{-i} \in U} u_i(x_i, x_{-i}) \geq \alpha_i > u^*_i$. This is a contradiction. Hence, x^* is a Nash equilibrium of G. ∎

The function $\bar{u}$ is like $\underline{u}$ except that players cannot "secure payoffs" using a (general) well-behaved correspondence. Instead, they must use a single strategy, or equivalently, a singleton-valued constant correspondence. This leads to the following definition. Let $G = (X_i, u_i)_{i \in N}$ be a normal-form game and $i \in N$. We say that u_i is *payoff secure* if, for all $x \in X$ and $\varepsilon > 0$, there exists $\bar{x}_i \in X_i$ and $U \in N(x_{-i})$ such that $u_i(\bar{x}_i, x'_{-i}) \geq u_i(x) - \varepsilon$ for all $x'_{-i} \in U$. Furthermore, G is *payoff secure* if u_i is payoff secure for all $i \in N$.

We refer to the function $\bar{u}$ as the *payoff secure envelope* of u. This terminology is justified by Lemma 3.11 below, which is analogous to Lemma 3.8.

Lemma 3.11 requires the following notation. Let $L^p(G)$ be the set of all bounded functions $\tilde{u} : X \to \mathbb{R}^n$ such that $\tilde{u}_i \leq u_i$ and $\tilde{u}_i$ is payoff secure for all $i \in N$.

Lemma 3.11   *Let $G = (X_i, u_i)_{i \in N} \in \mathbb{G}_q$. Then, for all $i \in N$:*

1. *For all $x \in X$, $u_i(x) \geq \bar{u}_i(x)$.*
2. *For all $x_{-i} \in X_{-i}$, $\bar{u}_i(\cdot, x_{-i})$ is quasiconcave.*
3. *The functions $\bar{u}_i(x_i, \cdot)$ and $\bar{v}_i$ are lower semicontinuous for all $x_i \in X_i$.*
4. *$\tilde{v}_i \leq \bar{v}_i$ for all $\tilde{u} \in L^p(G)$.*

Proof.   Let $i \in N$ and $x \in X$. For all $U \in N(x_{-i})$, we have that $x_{-i} \in U$. Thus, $\inf_{z_{-i} \in U} u_i(x_i, z_{-i}) \leq u_i(x)$ and so $\bar{u}_i(x) \leq u_i(x)$.

We turn to part 2. Let $\alpha \in \mathbb{R}$, $x_i, x'_i \in \{y_i \in X_i : \bar{u}_i(y_i, x_{-i}) > \alpha\}$ and $\lambda \in (0, 1)$. Then, there exist $U, U' \in N(x_{-i})$, such that $\inf_{z_{-i} \in U} u_i(x_i, z_{-i}) > \alpha$ and $\inf_{z_{-i} \in U'} u_i(x'_i, z_{-i}) > \alpha$.

Define $\bar{x}_i = \lambda x_i + (1 - \lambda)x'_i$ and $\bar{U} = U \cap U'$. It follows that $\bar{U}$ is an open neighborhood of x_{-i}. Let $z_{-i} \in \bar{U}$. The quasiconcavity of $u_i(\cdot, z_{-i})$

implies that

$$u_i(\bar{x}_i, z_{-i}) \geq \min\{u_i(x_i, z_{-i}), u_i(x'_i, z_{-i})\}$$

$$\geq \min\left\{\inf_{y_{-i} \in U} u_i(x_i, y_{-i}), \inf_{y_{-i} \in U'} u_i(x'_i, y_{-i})\right\}$$

and, thus, $\inf_{z_{-i} \in \bar{U}} u_i(\bar{x}_i, z_{-i}) > \alpha$. Hence, $\bar{u}_i(\bar{x}_i, x_{-i}) \geq \inf_{z_{-i} \in \bar{U}}$ $u_i(\bar{x}_i, z_{-i}) > \alpha$, which implies that $\bar{x}_i = \lambda x_i + (1 - \lambda)x'_i \in \{y_i \in X_i :$ $\bar{u}_i(y_i, x_{-i}) > \alpha\}$ and that $\bar{u}_i(\cdot, x_{-i})$ is quasiconcave.

We next show that $\bar{u}_i(x_i, \cdot)$ is lower semicontinuous for all $x_i \in X_i$. Let $x_i \in X_i$, $x_{-i} \in X_{-i}$ and $\varepsilon > 0$. Then, there exists $U \in N(x_{-i})$ such that $\inf_{z_{-i} \in U} u_i(x_i, z_{-i}) > \bar{u}_i(x) - \varepsilon$. We have that $U \in N(x'_{-i})$ for all $x'_{-i} \in U$. Thus, $\bar{u}_i(x_i, x'_{-i}) \geq \inf_{z_{-i} \in U} u_i(x_i, z_{-i}) > \bar{u}_i(x) - \varepsilon$ for all $x'_{-i} \in U$.

We next prove that $\bar{v}_i$ is lower semicontinuous. Since $\bar{u}_i$ is bounded, it follows that $\bar{v}_i$ is real-valued. Let $x_{-i} \in X_{-i}$ and $\varepsilon > 0$ be given. Let $0 < \eta < \varepsilon$ and let $x_i \in X_i$ be such that $\bar{u}_i(x_i, x_{-i}) - \eta > \bar{v}_i(x_{-i}) - \varepsilon$. Since $\bar{u}_i(x_i, \cdot)$ is lower semicontinuous, there exists an open neighborhood U of x_{-i} such that $\bar{u}_i(x_i, x'_{-i}) \geq \bar{u}_i(x) - \eta$ for all $x'_{-i} \in U$. Then, for all $x'_{-i} \in U$, we obtain that $\bar{v}_i(x'_{-i}) \geq \bar{u}_i(x_i, x'_{-i}) \geq \bar{u}_i(x) - \eta > \bar{v}_i(x_{-i}) - \varepsilon$. Hence, $\bar{v}_i$ is lower semicontinuous.

Finally, we show that $\bar{v}_i(x_{-i}) \geq \tilde{v}_i(x_{-i})$ for all $\tilde{u} \in L^p(G)$ and $x_{-i} \in X_{-i}$, where $\tilde{v}_i(x_{-i}) = \sup_{x_i \in X_i} \tilde{u}_i(x_i, x_{-i})$. Let $\tilde{u} \in L^p(G)$ and $x_{-i} \in X$ be fixed. Also, let $\varepsilon > 0$ and $x_i \in X_i$ be given. Since $\tilde{u} \in L^p(G)$, then there exists $U \in N(x_{-i})$ and $\bar{x}_i \in X_i$ such that $\tilde{u}_i(\bar{x}_i, z_{-i}) \geq \tilde{u}_i(x) - \varepsilon$ for all $z_{-i} \in U$. Since $u_i \geq \tilde{u}_i$, then

$$\bar{v}_i(x_{-i}) \geq \bar{u}_i(\bar{x}_i, x_{-i}) \geq \inf_{z_{-i} \in U} u_i(\bar{x}_i, z_{-i}) \geq \inf_{z_{-i} \in U} \tilde{u}_i(\bar{x}_i, z_{-i}) \geq \tilde{u}_i(x) - \varepsilon$$

and so $\bar{v}_i(x_{-i}) \geq \tilde{u}_i(x) - \varepsilon$. Since $\varepsilon > 0$ and x_i are arbitrary, then $\bar{v}_i(x_{-i}) \geq \tilde{v}_i(x_{-i})$. ∎

As a consequence of Lemma 3.11 we obtain the following characterization of better-reply security.

Theorem 3.12 *Let $G = (X_i, u_i)_{i \in N} \in \mathbb{G}_q$. Then the following conditions are equivalent:*

1. *G is weakly better-reply secure relative to some $\tilde{u} \in L^p(G)$.*
2. *G is weakly better-reply secure relative to $\bar{u}$.*
3. *G is better-reply secure.*

Proof.   Note first that Theorem 3.10 and Lemma 3.11 show that conditions 2 and 3 are equivalent. Lemma 3.11 also shows that $\bar{u} \in L^p(G)$. Hence, condition 2 implies condition 1.

Finally, suppose that condition 1 holds and let $\tilde{u} \in L^p(G)$ be such that G is weakly better-reply secure. Then $\bar{v}_i \geq \tilde{v}_i$ for all $i \in N$ and condition 3 follows.   ∎

We next consider the relationship between better-reply security and reducible security relative to $\bar{u}$. We obtain, surprisingly, that a result analogous to Theorem 3.9 does not hold in the case of better-reply security and reducible security relative to $\bar{u}$. This conclusion can be understood using the following characterization of better-reply security.

Theorem 3.13   *Let* $G = (X_i, u_i)_{i \in N} \in \mathbb{G}_q$. *Then,* G *is better-reply secure if and only if for all* $x \in E(G)^c$, *there exist* $\varepsilon > 0$ *and an open neighborhood* U *of* x *such that, for all* $x' \in U$, *there exists* $i \in N$ *with* $\bar{v}_i(x_{-i}) \geq u_i(x') + \varepsilon$.

Proof.   (Necessity) Suppose that G is such that there exists $x^* \notin E(G)$ such that, for all $\varepsilon > 0$ and $U \in N(x^*)$, there exists $x' \in U$ such that $\bar{v}_i(x^*_{-i}) < u_i(x') + \varepsilon$ for all $i \in N$. We will show that G is not generalized better-reply secure.

Let, for all $k \in \mathbb{N}$, $\varepsilon = 1/k$ and U be the open ball of radius $1/k$ around x^*. Then, by the above condition, we obtain a sequence $\{x_k\}_{k=1}^{\infty}$ such that $\lim_k x_k = x^*$ and $\bar{v}_i(x^*_{-i}) < u_i(x_k) + 1/k$ for all $i \in N$.

Since u is bounded, taking a subsequence if necessary, we may assume that $\{u(x_k)\}_{k=1}^{\infty}$ converges. Let $u^* = \lim_k u(x_k)$ and note that $(x^*, u^*) \in \text{cl}(\text{graph}(u))$. Since $\bar{v}_i(x^*_{-i}) < u_i(x_k) + 1/k$ for all $k \in \mathbb{N}$ and $i \in N$, then $\bar{v}_i(x^*_{-i}) \leq u_i^*$ for all $i \in N$. Since $x^* \notin E(G)$, this implies that G is not generalized better-reply secure.

(Sufficiency) Let $(x^*, u^*) \in \text{cl}(\text{graph}(u))$ be such that $x^* \notin E(G)$ and let $\{x_k\}_{k=1}^{\infty}$ be a sequence such that $\lim_k(x_k, u(x_k)) = (x^*, u^*)$. Also, let $\varepsilon > 0$ and $U \in N(x^*)$ be such that for all $x' \in U$ there exists $i \in N$ with $\bar{v}_i(x^*_{-i}) \geq u_i(x') + \varepsilon$. Since there are finitely many players and $x_k \in U$ for all sufficiently large k, there is $i \in N$ such that $\bar{v}_i(x^*_{-i}) \geq u_i(x_k) + \varepsilon$ for infinitely many k. Hence, $\bar{v}_i(x^*_{-i}) \geq u_i^* + \varepsilon > u_i^*$. Thus, G is better-reply secure by Theorem 3.10.   ∎

Theorem 3.13 shows that the difference between better-reply security and reducible security relative to $\bar{u}$ arises because the former uses u_i in the left-hand side of the inequality in the statement of Theorem 3.13 whereas the latter uses $\bar{u}_i$. Since $\bar{u} \le u$, then it follows that reducible security relative to $\bar{u}$ is weaker than better-reply security.

It is clear from Theorem 3.13 that reducible security relative to $\bar{u}$ coincides with better-reply security in games $G = (X_i, u_i)_{i \in N}$ such that $u = \bar{u}$. This happens precisely when $u_i(x_i, \cdot)$ is lower semicontinuous for all $i \in N$ and $x_i \in X_i$.

Theorem 3.14 *Let $G = (X_i, u_i)_{i \in N} \in \mathbb{G}_q$. Then, $u = \bar{u}$ if and only if $u_i(x_i, \cdot)$ is lower semicontinuous for all $i \in N$ and $x_i \in X_i$.*

Proof. (Necessity) Suppose that $u = \bar{u}$. Then $u_i(x_i, \cdot) = \bar{u}_i(x_i, \cdot)$ is lower semicontinuous by Lemma 3.11.

(Sufficiency) Suppose that $u_i(x_i, \cdot)$ is lower semicontinuous for all $i \in N$ and $x_i \in X_i$. Let $i \in N$, $x \in X$ and $\varepsilon > 0$. Since $u_i(x_i, \cdot)$ is lower semicontinuous, there exists an open neighborhood U of x_{-i} such that $u_i(x_i, x'_{-i}) > u_i(x) - \varepsilon$ for all $x_{-i} \in U$. Hence, $\bar{u}_i(x) \ge \inf_{x'_{-i} \in U} u_i(x_i, x'_{-i}) \ge u_i(x) - \varepsilon$. Since $u_i(x) \ge \bar{u}_i(x)$ by Lemma 3.11 and $\varepsilon > 0$ is arbitrary, it follows that $u_i(x) = \bar{u}_i(x)$. Thus, $u = \bar{u}$. ∎

The following result summarizes the relationship between better-reply security and reducible security relative to $\bar{u}$ discussed above. Furthermore, using an example, we show that reducible security relative to $\bar{u}$ is strictly weaker than better-reply security.

Theorem 3.15 *Let $G = (X_i, u_i)_{i \in N} \in \mathbb{G}_q$. Then the following holds:*

1. *If G is better-reply secure, then G is reducible secure relative to $\bar{u}$.*
2. *If G is reducible secure relative to $\bar{u}$ and is such that $u_i(x_i, \cdot)$ is lower semicontinuous for all $i \in N$ and $x_i \in X_i$, then G is better-reply secure.*

Proof. Parts 1 and 2 follow from Theorem 3.13 because $u_i \ge \bar{u}_i$ for all $i \in N$ (Lemma 3.11) and $u_i = \bar{u}_i$ when $u_i(x_i, \cdot)$ is lower semicontinuous for all $i \in N$ and $x_i \in X_i$ (Theorem 3.14). ∎

We now present an example of a normal-form game $G \in \mathbb{G}_q$ such that G is reducible secure relative to $\bar{u}$ but not better-reply secure. The game G is also payoff secure; thus, it is an example of game which is payoff secure and reducible secure relative to $\bar{u}$, but fails to be better-reply secure.

Example 3.16   Let G be such that $N = \{1,2\}$, $X_1 = X_2 = [0,1]$, and, for all $x \in X$,

$$u_1(x) = \begin{cases} 2 & \text{if } x_1 \geq 3/4, \\ 2 & \text{if } x_2 = 1 \text{ and } x_1 > 1/2, \\ 1 & \text{if } x_2 < 1 \text{ and } 1/2 < x_1 < 3/4, \\ 0 & \text{otherwise,} \end{cases}$$

and

$$u_2(x) = \begin{cases} 2 & \text{if } x_2 = 1, \\ 0 & \text{otherwise.} \end{cases}$$

It is easy to see that G is quasiconcave. Furthermore, G is payoff secure: for all $i \in N$, $x \in X$ and $\varepsilon > 0$, let $\bar{x}_i = 1$ and note that $u_i(\bar{x}_i, x'_{-i}) = 2 > u_i(x) - \varepsilon$ for all $x'_{-i} \in X_{-i}$.

We have that $E(G) = (1/2, 1] \times \{1\}$. Thus, $E(G)$ is not closed and, therefore, G is not better-reply secure (this follows by Theorems 4.10 and 4.12).

Finally, we show that G is reducible secure relative to $\bar{u}$. We have that

$$\bar{u}_1(x) = \begin{cases} 2 & \text{if } x_1 \geq 3/4, \\ 1 & \text{if } 1/2 < x_1 < 3/4, \\ 0 & \text{otherwise,} \end{cases}$$

$\bar{u}_2 = u_2$, $\underline{v}_1 \equiv 2$ and $\underline{v}_2 \equiv 2$.

Let $x \in E(G)^c$ and assume, first, that $x_2 \neq 1$. In this case, let $\varepsilon = 1$ and $U = \{x \in X : x_2 \neq 1\}$. Then, for all $x' \in U$, we have that $\underline{v}_1(x_1) = 2 \geq 1 = \bar{u}_2(x') + \varepsilon$.

Suppose next that $x_2 = 1$. Since $x \notin E(G)^c$, then $x_1 \leq 1/2$. Let $\varepsilon = 1$ and $U = \{x \in X : x_1 < 3/4\}$. Then, for all $x' \in U$, we have that $\underline{v}_1(x_2) = 2 \geq \bar{u}_1(x') + \varepsilon$. This shows that G is reducible secure relative to $\bar{u}$.

3.5   Sufficient Conditions

We next provide sufficient conditions for generalized better-reply security which are typically easier to verify. The conditions we will emphasize in this section are such that there is no need, in the games $G = (X_i, u_i)_{i \in N}$

where they hold, to replace players' payoff functions with $\underline{u}$, i.e., the game is weakly better-reply secure relative to u.

Note that a game $G = (X_i, u_i)_{i \in N} \in \mathbb{G}_q$ is weakly better-reply secure relative to u if u_i is generalized payoff secure for all $i \in N$ and x^* is a Nash equilibrium of G for all $(x^*, u^*) \in \Gamma$ such that $u_i^* \geq v_i(x_{-i}^*)$ for all $i \in N$. Thus, weak better-reply security relative to u combines generalized payoff security with another property which we will now call better-reply closedness relative to u. More generally, given a bounded function $\tilde{u} : X \to \mathbb{R}^n$, we say that G is *better-reply closed relative to* $\tilde{u}$ if $\tilde{u}_i \leq u_i$ and x^* is a Nash equilibrium of G for all $(x^*, u^*) \in \Gamma$ such that $u_i^* \geq \tilde{v}_i(x_{-i}^*)$ for all $i \in N$. Clearly, we have that G is weakly better-reply secure relative to u if and only if G is generalized payoff secure and better-reply closed relative to u.

The following result establishes the relationship between generalized better-reply security, better-reply closedness relative to u and weak better-reply security relative to u.

Theorem 3.17 *Let $G = (X_i, u_i)_{i \in N} \in \mathbb{G}_q$. Then:*

1. *If G is a generalized better-reply secure game, then G is better-reply closed relative to u.*
2. *If G is generalized payoff secure, then $u = \underline{u}$.*
3. *If G is better-reply closed relative to u and generalized payoff secure, then G is generalized better-reply secure.*

Proof. Part 1 follows because, by Lemma 3.4, $u_i \geq \underline{u}_i$ and, therefore, $v_i \geq \underline{v}_i$ for all $i \in N$.

Regarding part 2, if G is generalized payoff secure then $u \in L(G)$. Hence, Lemma 3.8 implies that $\underline{u} \geq u$. This, together with $u \geq \underline{u}$, implies that $\underline{u} = u$.

We finally establish part 3. If G is better-reply closed relative to u and generalized payoff secure, then G is weakly better-reply secure (relative to u) and, by Theorem 3.7, generalized better-reply secure. ∎

Theorem 3.18 below characterizes those games $G = (X_i, u_i)_{i \in N}$ that are weakly better-reply secure relative to u. Part of the characterization relies on the notion of weak reciprocal upper semicontinuity. A game $G = (X_i, u_i)_{i \in N}$ is *weakly reciprocal upper semicontinuous* if for all (x, α) in the frontier of the graph of u, there exists $i \in N$ and $\hat{x}_i \in X_i$ such that $u_i(\hat{x}_i, x_{-i}) > \alpha_i$. Moreover, for all $x \in X$, we say that G is *weakly reciprocal upper semicontinuous at x* if for all $\alpha \in \mathbb{R}^n$ such that (x, α) is in the frontier of the graph of u, there exists $i \in N$ and $\hat{x}_i \in X_i$ such that $u_i(\hat{x}_i, x_{-i}) > \alpha_i$.

Note that, clearly, G is weakly reciprocal upper semicontinuous if and only if it is weakly reciprocal upper semicontinuous at x for all $x \in X$.

Theorem 3.18   *Let $G = (X_i, u_i)_{i \in N} \in \mathbb{G}_q$. Then, G is better-reply closed relative to u if and only if G is weakly reciprocal upper semicontinuous at x^* for all $x^* \notin E(G)$.*

Consequently, G is weakly better-reply secure relative to u if and only if G is generalized payoff secure and weakly reciprocal upper semicontinuous at x^* for all $x^* \notin E(G)$.

Proof.   Suppose that G is better-reply closed relative to u. Let (x^*, u^*) be in the frontier of graph(u) and be such that x^* is not a Nash equilibrium of G. Then, in particular, $(x^*, u^*) \in \Gamma$ and better-reply closedness relative to u implies that there is $i \in N$ such that $v_i(x^*_{-i}) > u^*_i$. Hence, there is $x_i \in X_i$ such that $u_i(x_i, x^*_{-i}) > u^*_i$. Thus, G is weak reciprocally upper semicontinuous at x^*.

Conversely, suppose that G is weakly reciprocal upper semicontinuous at x^* for all $x^* \notin E(G)$. Let $(x^*, u^*) \in \Gamma$ be such that $u^*_i \geq v_i(x^*_{-i})$ for all $i \in N$. If x^* is not an equilibrium, then there is $j \in N$ such that $v_j(x^*_{-j}) > u_j(x^*)$ and so $u^*_j > u_j(x^*)$. This implies that (x^*, u^*) belongs to the frontier of graph(u). Since G is weak reciprocally upper semicontinuous at x^*, then there is $i \in N$ such that $u^*_i < v_i(x^*_{-i})$, a contradiction. Hence, x^* is a Nash equilibrium and G better-reply closed relative to u. ∎

An alternative characterization of better-reply closedness is obtained as follows. Given a normal-form game $G = (X_i, u_i)_{i \in N}$, define $\hat{u} : X \to \mathbb{R}^n$ by setting, for all $x \in X$,

$$\hat{u}(x) = \begin{cases} v(x) & \text{if there exists a sequence } \{x_k\}_{k=1}^{\infty} \text{ such that } \lim_k x_k = x \\ & \text{and } \lim_k u(x_k) \geq v(x), \\ u(x) & \text{otherwise} \end{cases}$$

We say that G is *regular* if $u = \hat{u}$.

Theorem 3.19   *Let $G = (X_i, u_i)_{i \in N} \in \mathbb{G}_q$. Then, G is better-reply closed relative to u if and only if G is regular.*

Proof.   (Necessity) Let $x \in X$ be such that there exists a sequence $\{x_k\}_{k=1}^{\infty}$ such that $\lim_k x_k = x$ and $\lim_k u(x_k) \geq v(x)$. Then $(x, \lim_k u(x_k)) \in \text{cl}(\text{graph}(u))$ and $\lim_k u(x_k) \geq v(x)$. Since G is better-reply closed relative to u, it follows that x is a Nash equilibrium of G. Thus, $u(x) = v(x)$. Since $\hat{u}(x) = v(x)$, it follows that $u(x) = \hat{u}(x)$ and G is regular.

(Sufficiency) Let $(x^*, u^*) \in \Gamma$ be such that $u_i^* \geq v_i(x_{-i}^*)$ for all $i \in N$. Then there exists a sequence $\{x_k\}_{k=1}^{\infty}$ such that $(x_k, u(x_k)) \to (x^*, u^*)$. Hence, $\lim_k u_i(x_k) = u_i^* \geq v_i(x_{-i}^*)$ for all $i \in N$ and so regularity implies that $u_i(x^*) = v_i(x_{-i}^*)$ for all $i \in N$. Thus, x^* is a Nash equilibrium and G is better-reply secure relative to u. ∎

The next result presents sufficient conditions for better-reply closedness. We say that a normal-form game $G = (X_i, u_i)_{i \in N}$ is *reciprocally upper semicontinuous* if $u(x) = u^*$ for all $(x^*, u^*) \in \mathrm{cl}(\mathrm{graph}(u))$ such that $u_i(x) \leq u_i^*$ for all $i \in N$. Furthermore, G is *sum-usc* if $x \mapsto \sum_{i \in N} u_i(x)$ is upper semicontinuous and is *upper semicontinuous* if u_i is upper semicontinuous for all $i \in N$.

Theorem 3.20 *Let $G = (X_i, u_i)_{i \in N} \in \mathbb{G}_q$. Then:*

1. *If G is weakly reciprocal upper semicontinuous then G is better-reply closed relative to u.*
2. *If G is reciprocal upper semicontinuous then G is weakly reciprocal upper semicontinuous.*
3. *If G is sum-usc then G is reciprocal upper semicontinuous.*
4. *If G is upper semicontinuous then G is sum-usc.*

Proof. Part 1 follows from Theorem 3.18. Regarding part 2, let (x^*, u^*) in the frontier of $\mathrm{graph}(u)$ and suppose that $u_i(x^*) \leq u_i^*$ for all $i \in N$. Since G is reciprocal upper semicontinuous, then $u(x^*) = u^*$, which implies that $(x^*, u^*) \in \mathrm{graph}(u)$, i.e., (x^*, u^*) does not belong to the frontier of $\mathrm{graph}(u)$. This contradiction establishes that there is $i \in N$ such that $u_i(x^*) > u_i^*$ and this shows that G is weakly reciprocal upper semicontinuous.

We next establish part 3. Let $(x^*, u^*) \in \mathrm{cl}(\mathrm{graph}(u))$ be such that $u_i(x^*) \leq u_i^*$ for all $i \in N$ and let $\{x_k\}_{k=1}^{\infty} \subseteq X$ be such that $\lim_k (x_k, u(x_k)) = (x^*, u^*)$. Since G is sum-usc then $\sum_{i=1}^{n} u_i(x^*) \geq \lim_k \sum_{i=1}^{n} u_i(x_k) = \sum_{i=1}^{n} u_i^*$. This, together with $u_i(x^*) \leq u_i^*$ for all $i \in N$, implies that $u_i(x^*) = u_i^*$ for all $i \in N$. Thus, G is reciprocal upper semicontinuous.

Finally, we prove part 4. Let $x \in X$ and $\{x_k\}_{k=1}^{\infty} \subseteq X$ be such that $\lim_k x_k = x$. Since G is upper semicontinuous, then $u_i(x) \geq \limsup_k u_i(x_k)$ for all $i \in N$. Since $\sum_{i=1}^{n} \limsup_k u_i(x_k) \geq \limsup_k \sum_{i=1}^{n} u_i(x_k)$, it follows that $\sum_{i=1}^{n} u_i(x) \geq \sum_{i=1}^{n} \limsup_k u_i(x_k) \geq \limsup_k \sum_{i=1}^{n} u_i(x_k)$. Hence, G is sum-usc. ∎

We next provide sufficient conditions for generalized payoff security.

Theorem 3.21   *Let $G = (X_i, u_i)_{i \in N} \in \mathbb{G}_q$. Then:*

1. *If G is payoff secure then G is generalized payoff secure.*
2. *If G is such that, for all $i \in N$ and $x_i \in X_i$, $u_i(x_i, \cdot)$ is lower semi-continuous, then G is payoff secure.*
3. *If G is lower semicontinuous, then $u_i(x_i, \cdot)$ is lower semicontinuous for all $i \in N$ and $x_i \in X_i$.*

Proof.   Parts 1 and 3 are obvious. Regarding Part 2, let $i \in N, x \in X$ and $\varepsilon > 0$. Since $u_i(x_i, \cdot)$ is lower semicontinuous, there exists an open neighborhood $V_{x_{-i}}$ of x_{-i} such that $u_i(x_i, x'_{-i}) > u_i(x_i, x_{-i}) - \varepsilon$ for all $x'_{-i} \in V_{x_{-i}}$. Thus, G is payoff secure.   ∎

The previous results can be interpreted as stating that generalized better-reply security combines weak forms of upper and lower semicontinuity. This is clearly seen by combining the weakest conditions in Theorems 3.20 and 3.21.

An alternative way of combining upper and lower semicontinuity conditions to yield an existence result appears in Dasgupta and Maskin (1986a, Corollary). In fact, this result shows that a Nash equilibrium exists in every game $G \in \mathbb{G}_q$ that is upper semicontinuous and weakly payoff secure. The latter condition is defined as follows: A game $G = (X_i, u_i)_{i \in N}$ is *weakly payoff secure* if v_i is lower semicontinuous for all $i \in N$. This condition is easily seen to be equivalent to the following weak form of payoff security: For all $i \in N$, $x \in X$ and $\varepsilon > 0$; there exists an open neighborhood U of x_{-i} with the following property: for all $x'_{-i} \in U$, there exists $\bar{x}_i \in X_i$ such that $u_i(\bar{x}_i, x'_{-i}) \geq u_i(x_i, x_{-i}) - \varepsilon$.

The above result and the combination of the weakest conditions in Theorems 3.20 and 3.21 give us two sets of sufficient conditions for the existence of equilibrium: (1) G is weakly reciprocal upper semicontinuous and generalized payoff secure and (2) G is upper semicontinuous and weakly payoff secure.

Comparing these two sets of conditions suggests that weakening the lower semicontinuity requirement from generalized payoff security to weak payoff security requires a strengthening of the upper semicontinuity one. We show that this is the case by presenting an example of a game that, while satisfying weak reciprocal upper semicontinuity and weak payoff security, fails to have a Nash equilibrium. In fact, the same conclusion holds even if we require the sum of players' payoff functions to be smooth.

Example 3.22   Let $G = (X_i, u_i)_{i \in N}$ be a two-player game with $X_1 = X_2 = [0, 1]$. Let $f : X \to \mathbb{R}$ be defined by $f(x) = 1 - (x_1 - x_2)^2$, $A =$

$\{x \in X : x_1 \geq x_2\}$, $B = \{x \in X : x_1 < x_2\}$, $C = \{(1,0)\}$ and, for every subset D of X, let χ_D denote the characteristic function of D. Define $u_1(x) = f(x)\chi_A(x) - \chi_C(x)$ and $u_2(x) = f(x)\chi_B(x) + \chi_C(x)$ for all $x \in X$. Thus, for all $x \in X$,

$$u_1(x) = \begin{cases} -1 & \text{if } x = (1,0), \\ 1 - (x_1 - x_2)^2 & \text{if } x_1 \geq x_2 \text{ and } x \neq (1,0), \\ 0 & \text{otherwise}, \end{cases}$$

and

$$u_2(x) = \begin{cases} 1 & \text{if } x = (1,0), \\ 1 - (x_1 - x_2)^2 & \text{if } x_1 < x_2, \\ 0 & \text{otherwise}. \end{cases}$$

Note that G is compact and quasiconcave. Moreover, the sum of players' payoff functions, $u_1 + u_2$, is equal to f and therefore smooth. Furthermore, $v_1(x_2) = v_2(x_1) = 1$ for all $x_1, x_2 \in [0,1]$ and so G is weakly payoff secure. However, G has no Nash equilibrium. In fact, the graph of player 1's beter-reply correspondence is the diagonal of X, whereas that of player 2 equals $C = \{(1,0)\}$, and so they do not intersect.

The above example shows that weak reciprocal upper semicontinuity needs to be strengthen in order for weakly payoff secure games to have Nash equilibria. Such a strengthening is as follows. Let $G = (X_i, u_i)_{i \in N}$ be a normal-form game and define $\breve{u} : X \times X \to \mathbb{R}^n$ by

$$\breve{u}(x, y) = (u_1(x_1, y_{-1}), \ldots, u_n(x_n, y_{-n}))$$

for all $(x, y) \in X \times X$. We say that G is *weakly upper semicontinuous* if, for all (x, y, α) in the frontier of the graph of $\breve{u}$, there exists $i \in N$ and $\hat{x}_i \in X_i$ such that $u_i(\hat{x}_i, y_{-i}) > \alpha_i$.

Theorem 3.23 below states that weak upper semicontinuity is indeed a weakening of upper semicontinuity and that it is a strengthening of weak reciprocal upper semicontinuity. Furthermore, it shows that weak payoff security is weaker than generalized payoff security.

Theorem 3.23 *Let $G = (X_i, u_i)_{i \in N} \in \mathbb{G}_q$.*

1. *If G is upper semicontinuous then G is weakly upper semicontinuous.*
2. *If G is weakly upper semicontinuous then G is weakly reciprocal upper semicontinuous.*
3. *If G is generalized payoff secure then G is weakly payoff secure.*

Proof.   Part 1: Let (x, y, u) be in the frontier of the graph of $\breve{u}$ and $i \in N$ be such that $u_i(x_i, y_{-i}) \neq u_i$. Let $\{(x_k, y_k)\}_{k=1}^{\infty}$ be such that $(x, y) = \lim_k (x_k, y_k)$ and $u = \lim_k \breve{u}(x_k, y_k)$. Then for all $i \in N$, $u_i = \lim_k u_i(x_i^k, y_{-i}^k)$ and, since player i's payoff function is upper semi-continuous, we obtain that $u_i(x_i, y_{-i}) \geq u_i$. This inequality combined with $u_i(x_i, y_{-i}) \neq u_i$ implies that $u_i(x_i, y_{-i}) > u_i$. Hence, simply let $\hat{x}_i = x_i$.

Part 2: Let (x, α) belong to the frontier of $\text{graph}(u)$ and let $\{x_k\}_{k=1}^{\infty} \subseteq X$ be such that $\lim_k (x_k, u(x_k)) = (x, \alpha)$. Then $\lim_k (x_k, x_k, \breve{u}(x_k, x_k)) = (x, x, \alpha)$ implying that $(x, x, \alpha) \in \text{cl}(\text{graph}(\breve{u}))$. Furthermore, $(x, x, \alpha) \in \text{graph}(\breve{u})$ if and only if $(x, \alpha) \in \text{graph}(u)$; hence, $(x, x, \alpha) \notin \text{graph}(\breve{u})$. Since G is weakly upper semicontinuous, then there exists $i \in N$ and $\hat{x}_i \in X_i$ such that $u_i(\hat{x}_i, x_{-i}) > \alpha$. This shows that G is weakly reciprocal upper semicontinuous.

Part 3: We have that $\underline{v}$ is lower semicontinuous by Lemma 3.4. Since G is generalized payoff secure, then $u = \underline{u}$ by Theorem 3.17 and thus $v = \underline{v}$ is lower semicontinuous. ∎

The following result provides two characterizations of weakly payoff secure and weakly upper semicontinuous games. The first characterization (i.e. the equivalence between the first two conditions in Theorem 3.24) shows that, when combined with weak payoff security, the requirement of weak upper semicontinuity becomes quite demanding. An alternative way of seeing this is by defining a function $\hat{u} : X^n \to \mathbb{R}^n$ by $\hat{u}(x^1, \ldots, x^n) = (u_1(x^1), \ldots, u_n(x^n))$ and noting that the second condition in Theorem 3.24 holds if and only if for all $(x^1, \ldots, x^n, u) \in \text{cl}(\text{graph}(\hat{u})) \backslash \text{graph}(\hat{u})$, there exists $i \in N$ and $\hat{x}_i \in X_i$ such that $u_i(\hat{x}_i, x_{-i}^i) > u_i$. Thus, the difference between weak reciprocal upper semicontinuity, weak upper semicontinuity imposed alone and weak reciprocal upper semicontinuity combined with weak payoff security is due to how the limit payoff vector is computed: for weak reciprocal upper semicontinuity, this vector is computed requiring all players to use the same strategy, whereas for weak upper semicontinuity this requirement is relaxed by allowing each player to change her own strategy; furthermore, when weak upper semicontinuity is combined with weak payoff security, the limit payoff vector is computed allowing different players to use different strategies.

The second characterization (i.e. the equivalence between the first and third conditions) implies that, in compact, weakly upper semicontinuous, weakly payoff-secure games, the best-reply correspondence is well-behaved. This equivalence also shows that, besides the continuity of players' value functions, weak upper semicontinuity and weak payoff security are almost necessary for the best-reply property to be well-behaved.

Theorem 3.24   *Let* $G = (X_i, u_i)_{i \in N} \in \mathbb{G}_q$. *Then the following conditions are equivalent*:

1. *G is weakly payoff secure and weakly upper semicontinuous.*
2. *G is weakly payoff secure and for all* $i \in N$ *and* $(x, \alpha) \in \mathrm{cl}(\mathrm{graph}(u_i)) \backslash \mathrm{graph}(u_i)$, *there exists* $\hat{x}_i \in X_i$ *such that* $u_i(\hat{x}_i, x_{-i}) > \alpha$.
3. *For all* $i \in N$, v_i *is continuous and the following property holds: if* $x \in X$, $\{x_k\}_{k=1}^{\infty} \subseteq X$ *and* $\{\varepsilon_k\}_{k=1}^{\infty} \subseteq \mathbb{R}_+$ *are such that* $\lim_k x_k = x$, $\lim_k \varepsilon_k = 0$ *and* $u_i(x_k) \geq v_i(x_{-i}^k) - \varepsilon_k$ *for all* $k \in \mathbb{N}$, *then* $u_i(x) = v_i(x_{-i})$.

Proof.   We start by establishing that 1 implies 2. Let $i \in N$ and $(x, \alpha_i) \in \mathrm{cl}(\mathrm{graph}(u_i)) \backslash \mathrm{graph}(u_i)$. If $u_i(x) > \alpha_i$, simply let $\hat{x}_i = x_i$; thus, we may assume that $u_i(x) < \alpha_i$. Let $\{x_k\}_{k=1}^{\infty}$ be such that $\lim_k x_k = x$ and $\lim_k u_i(x_k) = \alpha_i$. For all $k \in \mathbb{N}$, define $z_i^k = x_i^k$ and, for all $j \neq i$, let $z_j^k \in X_j$ be such that $u_j(z_j^k, x_{-j}^k) > v_j(x_{-j}^k) - 1/k$. Since G is compact, we may assume that $\{z_k\}_{k=1}^{\infty}$ and $\{v_j(x_{-j}^k)\}_{k=1}^{\infty}$ converge for all $j \neq i$. Let $z = \lim_k z_k$ and $\alpha_j = \lim_k v_j(x_{-j}^k)$ for all $j \neq i$. It follows that $(x, z, \alpha) = \lim_k(x_k, z_k, \breve{u}(z_k, x_k))$. Since G is weakly payoff secure, then $v_j(x_{-j}) \leq \alpha_j$ and so $u_j(\hat{x}_j, x_{-j}) \leq \alpha_j$ for all $j \neq i$ and $\hat{x}_j \in X_j$. Hence, by weak upper semicontinuity, there exists $\hat{x}_i \in X_i$ such that $u_i(\hat{x}_i, x_{-i}) > \alpha_i$.

We show next that 2 implies 3. Let $i \in N$, $x_{-i} \in X_{-i}$ and $\{x_{-i}^k\}_{k=1}^{\infty}$ be such that $\lim_k x_{-i}^k = x_{-i}$. Let $\{v_i(x_{-i}^m)\}_{m=1}^{\infty}$ be a subsequence of $\{v_i(x_{-i}^k)\}_{k=1}^{\infty}$. For all $m \in \mathbb{N}$, let $x_i^m \in X_i$ be such that $u_i(x_i^m, x_{-i}^m) > v_i(x_{-i}^m) - 1/m$. Since G is compact, then there is a subsequence $\{v_i(x_{-i}^n)\}_{n=1}^{\infty}$ of $\{v_i(x_{-i}^m)\}_{m=1}^{\infty}$ and a subsequence $\{x_i^n\}_{n=1}^{\infty}$ of $\{x_i^m\}_{m=1}^{\infty}$ such that both converge. By construction, $\lim_n u_i(x_n) = \lim_n v_i(x_{-i}^n)$ and so $(x, \lim_n v_i(x_{-i}^n)) \in \mathrm{cl}(\mathrm{graph}(u_i))$. Since v_i is lower semicontinuous by 2, we have that $u_i(x) \leq v_i(x_{-i}) \leq \lim_n v_i(x_{-i}^n)$. Thus, it follows again by 2 that $(x, \lim_n v_i(x_{-i}^n)) \in \mathrm{graph}(u_i)$; otherwise, for some $\hat{x}_i \in X_i$, $v_i(x_{-i}) \geq u_i(\hat{x}_i, x_{-i}) > \lim_n v_i(x_{-i}^n)$, a contradiction. Since $(x, \lim_n v_i(x_{-i}^n)) \in \mathrm{graph}(u_i)$, then $u_i(x) = \lim_n v_i(x_{-i}^n)$ and so $v_i(x_{-i}) = \lim_n v_i(x_{-i}^n)$. Since every subsequence of $\{v_i(x_{-i}^k)\}_{k=1}^{\infty}$ has a further subsequence converging to $v_i(x_{-i})$, we conclude that $v_i(x_{-i}) = \lim_k v_i(x_{-i}^k)$ and that v_i is continuous.

In order to establish the second condition in 3, consider $x \in X$, $\{x_k\}_{k=1}^{\infty} \subseteq X$ and $\{\varepsilon_k\}_{k=1}^{\infty} \subseteq \mathbb{R}_+$ such that $\lim_k x_k = x$, $\lim_k \varepsilon_k = 0$ and

$u_i(x_k) \geq v_i(x_{-i}^k) - \varepsilon_k$ for all $k \in \mathbb{N}$. Then, $\lim_k u_i(x_k) = \lim_k v_i(x_{-i}^k) = v_i(x_{-i})$. Thus, $(x, v_i(x_{-i})) \in \mathrm{cl}(\mathrm{graph}(u_i))$ and so $u_i(x) = v_i(x_{-i})$; otherwise, $(x, v_i(x_{-i}))$ belongs to the frontier of $\mathrm{graph}(u_i)$ and hence, for some $\hat{x}_i \in X_i$, $v_i(x_{-i}) \geq u_i(\hat{x}_i, x_{-i}) > v_i(x_{-i})$, a contradiction.

We finally establish that 3 implies 1. Since v_i is continuous for all $i \in N$, then G is weakly payoff secure. In order to see that G is weakly upper semicontinuous, consider (x, y, u) in the frontier of $\mathrm{graph}(\breve{u})$ and let $\{(x_k, y_k)\}_{k=1}^{\infty}$ be such that $(x, y) = \lim_k (x_k, y_k)$ and $u = \lim_k \breve{u}(x_k, y_k)$. If $u_i = v_i(y_{-i})$ for all $i \in N$, then, letting $\varepsilon_i^k = v_i(y_{-i}^k) - u_i(x_i^k, y_{-i}^k) \geq 0$ and $\varepsilon_k = \max_i \varepsilon_i^k$ for all $k \in \mathbb{N}$ and $i \in N$, we obtain $u_i(x_i^k, y_{-i}^k) \geq v_i(y_{-i}^k) - \varepsilon_k$, $\lim_k \varepsilon_k^i = v_i(y_{-i}) - u_i = 0$ for all $i \in N$ and so $\lim_k \varepsilon_k = 0$. It then follows by 3 that $u_i(x_i, y_{-i}) = v_i(y_{-i})$. Thus, $\breve{u}_i(x, y) = u_i(x_i, y_{-i}) = v_i(y_{-i}) = u_i$ and so $(x, y, u) \in \mathrm{graph}(\breve{u})$, a contradiction. Hence, there exists $i \in N$ such that $u_i \neq v_i(y_{-i})$. Since $u_i = \lim_k u_i(x_i^k, y_{-i}^k) \leq \lim_k v_i(y_{-i}^k) = v_i(y_{-i})$, then $u_i < v_i(y_{-i})$ and so there exists $\hat{x}_i \in X_i$ such that $u_i(\hat{x}_i, y_{-i}) > u_i$. Therefore, G is weakly upper semicontinuous. ∎

Theorem 3.24 implies, in particular, that in compact, quasiconcave, weakly upper semicontinuous and weakly payoff secure games, the best-reply correspondence is well-behaved. Furthermore, as we next show, it can be used to secure players' payoffs. This means that if $G \in \mathbb{G}_q$ is weakly upper semicontinuous and weakly payoff secure, then G is also generalized better-reply secure. Hence, the combination of upper and lower semicontinuity conditions in Carmona (2009) and in Dasgupta and Maskin (1986a, Corollary) are also particular ways of obtaining the generalized better-reply security of a game.

Theorem 3.25   *If $G = (X_i, u_i)_{i \in N} \in \mathbb{G}_q$ is weakly upper semicontinuous and weakly payoff secure then G is generalized better-reply secure.*

Proof.   First note that G is weakly reciprocal upper semicontinuous by Theorem 3.20. Thus, it suffices to show that G is generalized payoff secure. Let, for all $i \in N$, $B_i : X_{-i} \rightrightarrows X_i$ be player i's best-reply correspondence defined by $B_i(x_{-i}) = \{x_i \in X_i : u_i(x_i, x_{-i}) = v_i(x_{-i})\}$ for all $x_{-i} \in X_{-i}$.

Fix $i \in N$, $\varepsilon > 0$ and $x \in X$. Since v_i is lower semicontinuous, let $U \in N(x_{-i})$ be such that $v_i(x'_{-i}) > v_i(x_{-i}) - \varepsilon$ for all $x'_{-i} \in U$ and define $\varphi_i(x'_{-i}) = B_i(x'_{-i})$ for all $x'_{-i} \in U$. Then, by Theorem 3.24, φ_i is

a well-behaved correspondence and, for all $x' \in \text{graph}(\varphi_i)$, we have that
$u_i(x') = v_i(x'_{-i}) > v_i(x_{-i}) - \varepsilon \geq u_i(x) - \varepsilon.$ ∎

We conclude this section by providing sufficient conditions for weakly reciprocal upper semicontinuous and weakly payoff secure games to be generalized better-reply secure and, therefore, to have Nash equilibria. Theorem 3.27 below shows that such conclusion can be obtained by strengthening the quasiconcavity requirement. Theorem 3.26 provides a stronger result.

We say that a game $G = (X_i, u_i)_{i \in N}$ is *weakly continuous* if v_i is continuous for all $i \in N$. The following notion strengthens the usual quasiconcavity assumption. A game $G = (X_i, u_i)_{i \in N}$ is *locally joint quasiconcave* if for all $i \in N$ and $x \in X$, X_i is a polytope (and so a subset of an Euclidean space) and there exists an open neighborhood $V_{x_{-i}}$ of x_{-i} such that u_i is quasiconcave in $X_i \times V_{x_{-i}}$. The latter condition means that the set $\{x \in X_i \times V_{x_{-i}} : u_i(x) \geq \alpha\}$ is convex for all $\alpha \in \mathbb{R}$. Clearly, we then have that G is quasiconcave whenever G is locally joint quasiconcave.

Theorem 3.26 *If $G = (X_i, u_i)_{i \in N} \in \mathbb{G}_q$ is locally joint quasiconcave, better-reply closed relative to u and weakly continuous, then G is generalized better-reply secure.*

Proof. Note that it suffices to show that G is generalized payoff secure. Let $i \in N$, $\varepsilon > 0$ and $x \in X$. Let $0 < \eta < \varepsilon/2$ and, for all $\alpha > 0$, let $B_\alpha(x_{-i})$ denote the open ball of radius α around x_{-i} with respect to the sup norm of the Euclidean space containing X_{-i}. In particular, note that the closure of $B_\alpha(x_{-i})$ is a polytope.

Since v_i is continuous and u_i is locally quasiconcave, there is $\delta > 0$ such that u_i is quasiconcave in $X_i \times (X_{-i} \cap B_\delta(x_{-i}))$ and $|v_i(x'_{-i}) - v_i(\hat{x}_{-i})| < \eta$ for all $x'_{-i}, \hat{x}_{-i} \in X_{-i} \cap B_\delta(x_{-i})$. Let $U = X_{-i} \cap B_{\delta/2}(x_{-i})$ and note that $P := X_{-i} \cap \text{cl}(B_{\delta/2}(x_{-i}))$ is contained in $X_{-i} \cap B_\delta(x_{-i})$.

Since P is a polytope, let $x^0_{-i}, \ldots, x^m_{-i} \in P$ be such that $P = \text{co}(\{x^0_{-i}, \ldots, x^m_{-i}\})$. For convenience, let $\bar{x}_{-i} = (x^0_{-i}, \ldots, x^m_{-i})$. Consider the correspondence $\Theta_i : U \rightrightarrows \Delta_m$ defined by $\Theta_i(z_{-i}) = \{\theta \in \Delta_m : z_{-i} = \theta \cdot \bar{x}_{-i}\}$ for all $z_{-i} \in U$, where $\Delta_m = \{\theta \in \mathbb{R}^{m+1} : \sum_{j=0}^m \theta_j = 1$ and $\theta_j \geq 0$ for all $j\}$.

For all $j \in \{0, \ldots, m\}$, let $x^j_i \in X_i$ be such that $u_i(x^j_i, x^j_{-i}) > v_i(x^j_{-i}) - \varepsilon/2 + \eta$ and let $\bar{x}_i = (x^0_i, \ldots, x^m_i)$. Define $\varphi_i : U \rightrightarrows X_i$ by $\varphi_i(z_{-i}) = \{\theta \cdot \bar{x}_i : \theta \in \Theta_i(z_{-i})\}$ for all $z_{-i} \in U$. It is easy to show that φ_i is well-behaved.

We claim that $v_i(x'_{-i}) - \varepsilon/2 < u_i(x')$ for all $x' \in \operatorname{graph}(\varphi_i)$. If not, then $v_i(x'_{-i}) - \varepsilon/2 \geq u_i(x')$ for some $x' \in \operatorname{graph}(\varphi_i)$. Since $x'_i \in \varphi_i(x'_{-i})$, then $(x'_i, x'_{-i}) = \theta \cdot (\bar{x}_i, \bar{x}_{-i})$ for some $\theta \in \Delta_m$. The quasiconcavity of u_i in $X_i \times (X_{-i} \cap B_\delta(x_{-i}))$ implies that

$$u_i(x'_i, x'_{-i}) = u_i(\theta \cdot (\bar{x}_i, \bar{x}_{-i})) \geq \min_j u_i(x_i^j, x_{-i}^j) > \min_j v_i(x_{-i}^j) - \varepsilon/2 + \eta.$$

Hence, $v_i(x'_{-i}) > \min_j v_i(x_{-i}^j) + \eta$, a contradiction since $x'_{-i}, x_{-i}^j \in B_\delta(x_{-i})$ for all j.

Thus, $\varphi_i : U \rightrightarrows X_i$ is a well-behaved correspondence and, for all $x' \in \operatorname{graph}(\varphi_i)$, we have that $u_i(x') > v_i(x'_{-i}) - \varepsilon/2 > v_i(x_{-i}) - \varepsilon \geq u_i(x) - \varepsilon$. Hence, G is generalized payoff secure. ∎

The upper semicontinuity of each player's value function can be obtained by strengthening the notion of local joint quasiconcavity as follows. A game $G = (X_i, u_i)_{i \in N}$ is *strongly quasiconcave* if G is locally joint quasiconcave and for all $i \in N$ and $x_{-i} \in X_{-i}$, there exists an open neighborhood $V_{x_{-i}}$ of x_{-i} such that v_i is polyhedral quasiconcave in $V_{x_{-i}}$, i.e. for all $\alpha \in \mathbb{R}$, there exists a polytope P such that $\{x'_{-i} \in V_{x_{-i}} : v_i(x'_{-i}) \geq \alpha\} = P \cap V_{x_{-i}}$.

Theorem 3.27 *If $G = (X_i, u_i)_{i \in N} \in \mathbb{G}_q$ is strongly quasiconcave, weakly reciprocal upper semicontinuous and weakly payoff secure, then G is generalized better-reply secure.*

Proof.   Note that if G is strongly quasiconcave, then v_i is upper semicontinuous for all $i \in N$. Indeed, consider $x_{-i} \in X_{-i}$ and $\{x_{-i}^k\}_{k=1}^\infty \subseteq X_{-i}$ such that $\lim_k x_{-i}^k = x_{-i}$. Let $\alpha = \limsup_k v_i(x_{-i}^k)$ and, taking a subsequence if needed, assume that $\lim_k v_i(x_{-i}^k) = \alpha$. Fix $\varepsilon > 0$ and let $V_{x_{-i}}$ and P be such that $V_{x_{-i}}$ is an open neighborhood of x_{-i}, P is a polytope and $\{x'_{-i} \in V_{x_{-i}} : v_i(x'_{-i}) \geq \alpha - \varepsilon\} = P \cap V_{x_{-i}}$. Since $x_{-i}^k \in \{x'_{-i} \in V_{x_{-i}} : v_i(x'_{-i}) \geq \alpha - \varepsilon\}$ for all k sufficiently large, then $x_{-i}^k \in P$ for all k sufficiently large. Since polytopes are compact, then $x_{-i} \in P \cap V_{x_{-i}} = \{x'_{-i} \in V_{x_{-i}} : v_i(x'_{-i}) \geq \alpha - \varepsilon\}$. Hence, $v_i(x_{-i}) \geq \alpha - \varepsilon$. Since $\varepsilon > 0$ is arbitrary, it follows that $v_i(x_{-i}) \geq \alpha = \limsup_k v_i(x_{-i}^k)$. This shows that v_i is upper semicontinuous.

Thus, G is weakly continuous and, moreover, locally joint quasiconcave and better-reply closed relative to u (by Theorem 3.20). Hence, G is generalized better-reply secure by Theorem 3.26. ∎

3.6 Mixed Strategies

We have seen before that non-quasiconcave games may fail to have Nash equilibria. In this case, one can seek the existence of mixed strategy Nash equilibria, i.e. of pure strategy Nash equilibria of the mixed extension of the game. Because a pure strategy of a game is a particular case of a mixed strategy, it follows that the strategy space of the mixed extension is larger than that of the original game; this makes it harder to analyze the latter, in particular, to check whether or not it is generalized better-reply secure. For this reason, it is useful to have sufficient conditions on the original game for the mixed extension to be generalized better-reply secure.

The following result gives sufficient conditions for the mixed extension to be sum-usc, which, by Theorems 3.18 and 3.20, implies that the mixed extension is better-reply closed.

Theorem 3.28   *Let $G = (X_i, u_i)_{i \in N} \in \mathbb{G}$. If G is sum-usc then $\bar{G}$ is also sum-usc.*

Proof.   Let $f : X \to \mathbb{R}$ be defined by $f(x) = \sum_{i=1}^{n} u_i(x)$. Since G is sum-usc, then f is upper semicontinuous. By Theorem A.23, $\mu \mapsto \int f \mathrm{d}\mu$ is upper semicontinuous. Since $\int f \mathrm{d}\mu = \sum_{i=1}^{n} \int u_i \mathrm{d}\mu = \sum_{i=1}^{n} u_i(\mu)$ for all $\mu \in M(X)$, then it follows that $\bar{G}$ is sum-usc.   ∎

The following example answers the question of whether or not the payoff security of the mixed extension $\overline{G}$ of a normal form game G follows from the payoff security of G. It shows that a game can be payoff secure in pure strategies without being payoff secure in mixed strategies.

Example 3.29   Let G be defined by $N = \{1, 2\}$, $X_1 = X_2 = [0, 1]$, $u_1 : X \to \mathbb{R}$ be defined by

$$
u_1(x_1, x_2) = \begin{cases} -1 & \text{if } x_1 < x_2 < x_1 + \dfrac{1}{2}, \\ 0 & \text{if } x_1 = x_2 \text{ or } x_2 = x_1 + \dfrac{1}{2}, \\ 1 & \text{otherwise}, \end{cases}
$$

and $u_2 : X \to \mathbb{R}$ be defined by $u_2(x_1, x_2) = -u_1(x_1, x_2)$, for all $(x_1, x_2) \in X$.

We first show that G is payoff secure. For player 1, it is enough to consider $x \in X$ such that $u_1(x) = 0$ since when $u_1(x) = -1$ there is nothing to show, and u_1 is continuous in the remaining case (i.e., for x such that $u_1(x) = 1$). If $x_1 = x_2$, we let $\bar{x}_1 = 1$; if $x_2 = x_1 + 1/2$, we let

$\bar{x}_1 = 0$ except when $x = (0, 1/2)$, in which case we let $\bar{x}_1 = 1$. In all these cases we can find a neighborhood V_{x_2} of x_2 such that $u_1(\bar{x}_1, x_2') \geq 0$ for all $x_2' \in V_{x_2}$.

For player 2, it is again enough to consider $x \in X$ such that $u_1(x) = 0$. We let $\bar{x}_2 = \min\{x_1 + 1/4, 1\}$. In all these cases we can find a neighborhood V_{x_1} of x_1 such that $u_2(x_1', \bar{x}_2) \geq 0$ for all $x_1' \in V_{x_1}$.

We next show that $\overline{G}$ is not payoff secure. For all $i \in N$ and $x_i \in X_i$, let $\delta_{x_i} \in M(X_i)$ be such that $\delta_{x_i}(\{x_i\}) = 1$. Let $\lambda = (\delta_0, \frac{1}{3}\delta_{1/2} + \frac{2}{3}\delta_1)$ and $0 < \varepsilon < 1/3$. We will show that, for all $\mu_1 \in M(X_1)$, there exists $\{\lambda_2^k\}_{k=1}^{\infty} \subseteq M(X_2)$ such that $\lim_k \lambda_2^k = \lambda_2$ and $u_1(\mu_1, \lambda_2^k) < u_1(\lambda) - \varepsilon$ for all k sufficiently large, which clearly implies that $\overline{G}$ is not payoff secure.

First, note that $u_1(\lambda) = u_1(0, 1/2)/3 + 2u_1(0, 1)/3 = 2/3$. Define, for all $k \in \mathbb{N}$, $\lambda_2^k = \frac{1}{3}\delta_{1/2 - 1/(2k)} + \frac{2}{3}\delta_1$. Since $\lim_k(1/2 - 1/(2k)) = 1/2$, it follows that $\lim_k \lambda_2^k = \lambda_2$. For all $\mu_1 \in M(X_1)$, we have that

$$u_1(\mu_1, \lambda_2^k) = \frac{1}{3}u_1(\mu_1, 1/2 - 1/(2k)) + \frac{2}{3}u_1(\mu_1, 1)$$

$$= \frac{1}{3}\left(-\mu_1\left(\left[0, \frac{1}{2} - \frac{1}{2k}\right)\right) + \mu_1\left(\left(\frac{1}{2} - \frac{1}{2k}, 1\right]\right)\right)$$

$$+ \frac{2}{3}\left(\mu_1\left(\left[0, \frac{1}{2}\right)\right) - \mu_1\left(\left(\frac{1}{2}, 1\right]\right)\right)$$

Since $\{1/2 - 1/(2k)\}_k$ increases to $1/2$, then

$$\lim_k u_1(\mu_1, \lambda_2^k) = \frac{1}{3}\left(-\mu_1\left(\left[0, \frac{1}{2}\right)\right) + \mu_1\left(\left[\frac{1}{2}, 1\right]\right)\right)$$

$$+ \frac{2}{3}\left(\mu_1\left(\left[0, \frac{1}{2}\right)\right) - \mu_1\left(\left(\frac{1}{2}, 1\right]\right)\right)$$

$$= \frac{1}{3}\mu\left(\left[0, \frac{1}{2}\right)\right) + \frac{1}{3}\mu\left(\left\{\frac{1}{2}\right\}\right) - \mu\left(\left[\frac{1}{2}, 1\right]\right)\right) + \frac{1}{3}\mu(\{1\})$$

$$\leq \frac{1}{3} < \frac{2}{3} - \varepsilon.$$

Thus, $u_1(\mu_1, \lambda_2^k) \leq 2/3 - \varepsilon$ for all k sufficiently large.

Example 3.29 has originally considered in Sion and Wolfe (1957), where it is shown that G has no Nash equilibrium (pure or mixed). Hence, G is also an example of a (non-quasiconcave) reciprocally upper semicontinuous (since G is a zero-sum game so is $\overline{G}$) and payoff secure game without Nash equilibria. Furthermore, this example shows that mixed extension of a better-reply secure game may fail to be better-reply secure (or even generalized better-reply secure).

The above example implies that a condition stronger than the payoff security of a game is needed for its mixed extension to be payoff secure. The following is one such condition. A normal-form game $G = (X_i, u_i)_{i \in N}$ is *uniform payoff secure* if for all $i \in N$, $x_i \in X_i$ and $\varepsilon > 0$, there exists $\bar{x}_i \in X_i$ such that for all $y_{-i} \in X_{-i}$ there exists a neighborhood $V_{y_{-i}}$ of y_{-i} such that $u_i(\bar{x}_i, y'_{-i}) \geq u_i(x_i, y_{-i}) - \varepsilon$ for all $y'_{-i} \in V_{y_{-i}}$.

The relationship between uniform payoff security and other conditions of the lower semicontinuity type is established next.

Theorem 3.30   *Let* $G = (X_i, u_i)_{i \in N} \in \mathbb{G}$. *Then:*

1. *If G is uniformly payoff secure then G is payoff secure.*
2. *If G is such that $u_i(x_i, \cdot)$ is lower semicontinuous for all $i \in N$ and $x_i \in X_i$, then G is uniformly payoff secure.*

Proof.   Part 1 is obvious. Regarding part 2, given $i \in N$, $x_i \in X_i$ and $\varepsilon > 0$, simply let $\bar{x}_i = x_i$. Then the lower semicontinuity of $u_i(x_i, \cdot)$ gives, for all $y_{-i} \in X_{-i}$, $V_{y_{-i}}$ with the desired properties. ∎

Theorem 3.31 shows that the uniform payoff security of a game G is indeed sufficient for $\overline{G}$ to be payoff secure.

Theorem 3.31   *If $G = (X_i, u_i)_{i \in N} \in \mathbb{G}$ is uniform payoff secure then $\overline{G}$ is payoff secure.*

Proof.   Let $i \in N$, $m \in M$ and $\varepsilon > 0$. Since m_i is a probability measure, there exists $\tilde{x}_i \in X_i$ such that

$$u_i(\tilde{x}_i, m_{-i}) \geq u_i(m). \tag{3.2}$$

Since G is uniformly payoff secure, there exists $\bar{x}_i \in X_i$ such that for all $y_{-i} \in X_{-i}$ there exists a neighborhood $V_{y_{-i}}$ of y_{-i} such that $u_i(\bar{x}_i, y'_{-i}) \geq u_i(\tilde{x}_i, y_{-i}) - \varepsilon$ for all $y'_{-i} \in V_{y_{-i}}$. This implies that

$$\bar{u}_i(\bar{x}_i, y_{-i}) \geq u_i(\tilde{x}_i, y_{-i}) - \varepsilon/2 \text{ for all } y_{-i} \in X_{-i}. \tag{3.3}$$

We have that $\bar{u}_i(\bar{x}_i, \cdot)$ is lower semicontinuous by Lemma 3.11. Hence, $\mu_{-i} \mapsto \int_{X_{-i}} \bar{u}_i(\bar{x}_i, x_{-i}) \mathrm{d}\mu_{-i}(x_{-i})$ is lower semicontinuous by Theorem A.23

and, therefore, there exists an open neighborhood U of m_{-i} such that

$$\bar{u}_i(\bar{x}_i, \mu_{-i}) > \bar{u}_i(\bar{x}_i, m_{-i}) - \varepsilon/2 \quad \text{for all } \mu_{-i} \in U. \tag{3.4}$$

Recall that $u_i \geq \bar{u}_i$ by Lemma 3.11. Then, for all $\mu_{-i} \in U$,

$$u_i(\bar{x}_i, \mu_{-i}) \geq \bar{u}_i(\bar{x}_i, \mu_{-i}) > \bar{u}_i(\bar{x}_i, m_{-i}) - \varepsilon/2 > u_i(\tilde{x}_i, m_{-i}) - \varepsilon$$

$$\geq u_i(m) - \varepsilon,$$

where the first inequality follows from $u_i \geq \bar{u}_i$, the second by (3.4), the third by (3.3) and the last by (3.2). ∎

Combining Theorems 3.28 and 3.31, we obtain sufficient conditions on pure strategies for the mixed extension of a game to be generalized better-reply secure.

Theorem 3.32   *If $G = (X_i, u_i)_{i \in N} \in \mathbb{G}$ is sum-usc and uniform payoff secure, then $\overline{G}$ is better-reply secure.*

Theorem 3.32 provides conditions on a game that imply that its mixed extension is generalized better-reply secure. However, sometimes it is easy to show that the mixed extension of a game is better-reply secure by analyzing the mixed extension directly.

The following result provides an example of such situation for the case of payoff security. It shows that the mixed extension of a game is payoff secure if each player can, through her choice and given the choice of the other players, avoid discontinuities while roughly maintaining her starting payoff.

Theorem 3.33   *Let $G = (X_i, u_i)_{i \in N} \in \mathbb{G}$ and suppose that the following property holds: For all $i \in N$, $\varepsilon > 0$, $x_i \in X_i$ and $m_{-i} \in M_{-i}$, there exists $\bar{x}_i \in X_i$ such that*

$$m_{-i}(\{x_{-i} \in X_{-i} : u_i \text{ is discontinuous at } (\bar{x}_i, x_{-i})\}) = 0$$

and $u_i(\bar{x}_i, m_{-i}) \geq u_i(x_i, m_{-i}) - \varepsilon$. Then $\overline{G}$ is payoff secure.

Proof.   Let $i \in N$, $m \in M$ and $\varepsilon > 0$. Since m_i is a probability measure, then there exists $x_i \in X_i$ such that $u_i(x_i, m_{-i}) \geq u_i(m)$. Furthermore, by assumption, there exists $\bar{x}_i \in X_i$ such that $m_{-i}(\{x_{-i} \in X_{-i} : u_i \text{ is discontinuous at } (\bar{x}_i, x_{-i})\}) = 0$ and $u_i(\bar{x}_i, m_{-i}) \geq u_i(x_i, m_{-i}) - \varepsilon/2$.

Since $m_{-i}(\{x_{-i} \in X_{-i} : u_i \text{ is discontinuous at } (\bar{x}_i, x_{-i})\}) = 0$, it follows by Theorem A.23 that $m' \mapsto u_i(m')$ is continuous at $(\bar{x}_i, m_{-i})$

and, hence, that $m'_{-i} \mapsto u_i(\bar{x}_i, m'_{-i})$ is continuous at m_{-i}. It then follows that there exists an open neighborhood V of m_{-i} such that $u_i(\bar{x}_i, m'_{-i}) > u_i(\bar{x}_i, m_{-i}) - \varepsilon/2$ for all $m'_{-i} \in V$. Hence, for all $m'_{-i} \in V$, $u_i(\bar{x}_i, m'_{-i}) > u_i(m) - \varepsilon$. ∎

3.7 References

The notion of generalized better-reply security is due to Barelli and Soza (2010), who have first established Theorem 3.2. Its proof, in particular Theorem 3.3 and Lemmas 3.4–3.6, is due to Carmona (2011c).

The first example in Section 3.2 is a slight modification of Example 2.2 in Barelli and Soza (2010). The second example in that section is due to Reny (1999).

Theorem 3.7 is due to Carmona (2011c) and Lemma 3.8 improves upon Lemma 4 of that paper. The notion of better-reply security is due to Reny (1999), while Theorems 3.10 and 3.12 and Lemma 3.11 are based on Carmona (2011c). Theorems 3.9 and 3.13–3.15 and Example 3.16 are based on Carmona (2011a).

Theorem 3.18 is due to Carmona (2011c). The notion of a regular game and Theorem 3.19 are due to Castro (2011). Bagh and Jofre (2006) introduced the concept of weak reciprocal upper semicontinuity and established part 2 of Theorem 3.20. The notion of sum-usc is due to Dasgupta and Maskin (1986a), while that of reciprocal upper semicontinuity is due to Simon (1987) and Reny (1999). Reny (1999) introduced the concept of payoff security, which was extended to generalized payoff security by Barelli and Soza (2010). Weak payoff security was first considered in Dasgupta and Maskin (1986a) and weak upper semicontinuity in Carmona (2009). Carmona (2009) also establishes Example 3.22, part 1 of Theorem 3.23 and Theorem 3.24. Theorem 3.25 is due to Barelli and Soza (2010). The remaining concepts and results in Section 3.5, weak continuity, local joint quasiconcavity, strong quasiconcavity and Theorems 3.26 and 3.27, are due to Carmona (2011c).

Finally, on Section 3.6, Theorem 3.6 is due to Reny (1999). Example 3.29 is due to Sion and Wolfe (1957) and its analysis to Carmona (2005) (pure strategies) and Monteiro and Page (2007) (mixed strategies). Carbonell-Nicolau and Ok (2007) and Monteiro and Page (2007) introduced the notion of uniform payoff security and Monteiro and Page (2007) established Theorem 3.31. The condition in Theorem 3.33 is due to Bagh (2010).

Chapter 4

Stronger Existence Results

In this chapter, we present several recent existence results. These results consider conditions on (compact, quasiconcave) games that are variations and extensions of better-reply security.

Recall that generalized better-reply security imposes a condition only on strategy profiles that are not Nash equilibria of the game in question. Moreover, this condition involves a regularized payoff function (the function $\underline{u}$ in a game $G = (X_i, u_i)_{i \in N}$) and a particular way of relating u and $\underline{u}$.

We consider five conditions in this chapter: Multi-player well-behaved security, diagonal transfer continuity, generalized C-security, the lower single-deviation property and generalized weak transfer continuity. All these conditions are, like generalized better-reply security, only imposed on strategy profiles that are not Nash equilibria. They differ from generalized better-reply security because they consider different regularized payoff function or different ways of relating the original and the regularized payoff functions. Furthermore, in some cases, they consider a property that generalized better-reply security satisfies.

We show that (compact, metric, quasiconcave) games satisfying any of the above conditions have a Nash equilibrium. Furthermore, we establish some relationship between the several conditions. We present an additional characterization of generalized better-reply security which, in particular, implies that every generalized better-reply secure game is generalized C-secure. Moreover, we also show that when a game is either diagonally transfer continuous or generalized C-secure on some compact subset of the strategy space, then such game is also multi-player well-behaved secure on that compact subset of strategies.

4.1   Multi-Player Well-Behaved Security

In this section we consider multi-player well-behaved secure games. These are games G with the property that there exists, for every strategy outside the set of Nash equilibria of G, a well-behaved correspondence defined locally and whose graph is contained in the complement of the set of Nash equilibria of G.

The formal definition of multi-player well-behaved security is as presented in what follows. First, we need the notion of a product correspondence. Let Y be a subset of $X = \prod_{i \in N} X_i$ and $\Phi : Y \rightrightarrows X$ be a correspondence. We say that Φ is a *product correspondence* if, for all $x \in Y$, $\Phi(x) = \prod_{i \in N} \Phi_i(x)$.

Let $G = (X_i, u_i)_{i \in N}$ be a normal-form game. For all $i \in N$ and $x \in X$, let $U_i(x) = \{y_i \in X_i : u_i(y_i, x_{-i}) > u_i(x)\}$. For all $Y \subseteq X$, we say that G is *multi-player well-behaved secure on Y* if there exists a family $\{V_x, \varphi_x\}_{x \in Y}$ such that

(a) V_x is an open neighborhood of x for all $x \in Y$,

(b) $\varphi_x : V_x \rightrightarrows X$ is a well-behaved product correspondence for all $x \in Y$, and

(c) for all $x \in Y$ and $y \in V_x$ there is $i \in N$ such that $\varphi_{z,i}(y) \in U_i(y)$ for all $z \in Y$ with $y \in V_z$.

A normal-form game $G \in \mathbb{G}$ is *multi-player well-behaved secure* if it is multi-player well-behaved secure on $E(G)^c$.

As the next result shows, multi-player well-behaved security is sufficient for the existence of Nash equilibria.

Theorem 4.1   *Let $G = (X_i, u_i)_{i \in N} \in \mathbb{G}_q$. If G is multi-player well-behaved secure then G has a Nash equilibrium.*

The following lemma presents the key technical result needed to establish Theorem 4.1.

Lemma 4.2   *Let $G = (X_i, u_i)_{i \in N} \in \mathbb{G}_q$ and K be a compact subset of X. If G is multi-player well-behaved secure on K, then there exists a well-behaved product correspondence $\Phi : K \rightrightarrows X$ such that, for all $x \in K$, there exist $i \in N$ such that $\Phi_i(x) \subseteq U_i(x)$.*

    Proof. Let $\{V_x, \varphi_x\}_{x \in K}$ be as in the definition of multi-player well-behaved security. Since K is compact, there exists a finite open cover $\{V_{x_j}\}_{j=1}^m$ of K and, by Theorem A.3, there exists a partition of unity $\{\beta_j\}_{j=1}^m$ subordinate to $\{V_{x_j}\}_{j=1}^m$. Define $\Phi : K \rightrightarrows X$ by $\Phi(x) = \sum_{j=1}^m \beta_j(x)\varphi_{x_j}(x)$. It follows by Theorem A.8 that Φ is a well-behaved correspondence. Furthermore, since φ_{x_j} is a product correspondence for all $j \in \{1, \ldots, m\}$, then Φ is also a product correspondence.

Let $x \in K$. Using part (c) of the definition of multi-player well-behaved security with $y = x$, it follows, in particular, that there is $i \in N$ such that $\varphi_{x_j}(x) \in U_i(x)$ for all $j \in \{1, \ldots, m\}$ such that $x \in V_{x_j}$. Since, for all j, $\beta_j(x) = 0$ if $x \notin V_{x_j}$ and since $U_i(x)$ is convex (due to $G \in \mathbb{G}_q$), it follows that $\Phi_i(x) = \sum_{j=1}^m \beta_j(x)\varphi_{x_j,i}(x) \in U_i(x)$. ∎

Theorem 4.1 can be established in a simple way with the help of Lemma 4.2. Suppose, in order to reach a contradiction, that $E(G) = \emptyset$. This implies that $E(G)^c = X$ and that G is multi-player well-behaved secure on X. Hence, by Lemma 4.2, there exists a well-behaved product correspondence $\Phi : X \rightrightarrows X$ such that for all $x \in X$ there exist $i \in N$ such that $\Phi_i(x) \subseteq U_i(x)$.

We next define a new game, $G_0' = (Z_i, r_i')_{i \in N_0}$ by setting, $N_0 = N \cup \{0\}$, $Z_0 = X$, $Z_i = X_i$ for all $i \in N$,

$$r_0'(z) = \begin{cases} 1 & \text{if } z_0 = z_{-0}, \\ 0 & \text{otherwise}, \end{cases}$$

for all $z \in Z = Z_0 \times Z_1 \times \cdots \times Z_n$ and, for all $i \in N$ and $z \in Z$,

$$r_i'(z) = \begin{cases} 1 & \text{if } z_i \in \Phi_i(z_0), \\ 0 & \text{otherwise} \end{cases}$$

The key properties of the game G_0' are: (1) $E(G_0') = \emptyset$ and (2) the best-reply correspondence of G_0' is well-behaved. But (1) and (2) together contradict the Cauty's fixed point theorem (Theorem A.14).

Proof of Theorem 4.1. Let Φ and G_0' be as above, and let B be the best-reply correspondence of G_0'. Clearly, we have that $B(z) = \{z_{-0}\} \times \Phi(z_0)$ for all $z \in K$. Thus, B is well-behaved.

We next show that $E(G_0') = \emptyset$. Suppose that $z \in E(G_0')$. Since $B(z) = \{z_{-0}\} \times \Phi(z_0)$, then $z_{-0} \in \Phi(z_0)$ and $z_0 = z_{-0}$. Letting $x = z_{-0}$, we have that $x \in \Phi(x)$ and, hence, there exists $i \in N$ such that $x_i \in U_i(x)$. But this is a contradiction because $x_i \in U_i(x)$ implies that $u_i(x) < u_i(x)$. Thus, $E(G_0') = \emptyset$. ∎

Multi-player well-behaved security not only guarantees the existence of Nash equilibria but it also implies that the set of Nash equilibria is closed.

Theorem 4.3   *If $G = (X_i, u_i)_{i \in N} \in \mathbb{G}$ is multi-player well-behaved secure then $E(G)$ is closed.*

Proof. Let $x \in E(G)^c$ and let V_x and φ_x be as in the definition of multi-player well-behaved security. Then, for all $y \in V_x$, there exists $i \in N$ such that $\varphi_{x,i}(y) \in U_i(y)$, i.e., $u_i(z_i, y_{-i}) > u_i(y)$ for all $z_i \in \varphi_{x,i}(y)$. Hence, $y \in E(G)^c$ and $V_x \subseteq E(G)^c$. It follows that $E(G)^c$ is open and that $E(G)$ is closed. ∎

4.2   Diagonal Transfer Continuity

We consider in this section the notion of diagonal transfer continuity. Let $G = (X_i, u_i)_{i \in N}$ be a normal-form game and define $U : X \times X \to \mathbb{R}$ by $U(x, y) = \sum_{i=1}^{n} u_i(x_i, y_{-i})$ for all $x, y \in X$. We say that G is *diagonally transfer continuous* if for all $(x, y) \in X \times X$ such that $U(x, y) > U(y, y)$ there exists $x' \in X$ and an open neighborhood V_y of y such that $U(x', z) > U(z, z)$ for all $z \in V_y$.

The function U can be used to characterize the Nash equilibria of G. This is so because $y \in X$ is a Nash equilibrium of G if and only if $U(x, y) \leq U(y, y)$ for all $x \in X$. Thus, $(x, y) \in X \times X$ is such that $U(x, y) > U(y, y)$ if and only if y is not a Nash equilibrium of G. As a consequence, we can rewrite the definition of diagonal transfer continuity as follows: G is diagonally transfer continuous if for all $y \in E(G)^c$ there exists $x' \in X$ and an open neighborhood V_y of y such that $U(x', z) > U(z, z)$ for all $z \in V_y$.

More generally, given $Y \subseteq X$, we say that G is *diagonally transfer continuous on Y* if for all $y \in Y$ there exists $x' \in X$ and an open neighborhood V_y of y such that $U(x', z) > U(z, z)$ for all $z \in V_y$.

The relationship between diagonal transfer continuity and multi-player well-behaved security is established in the next result. It shows that if a game is diagonally transfer continuous on a compact set, then it is also multi-player well-behaved secure on that compact set. However, this result require a stronger convexity assumption: We say that a normal-form game G is *concave* if $u_i(\cdot, x_{-i})$ is concave for all $x_{-i} \in X_{-i}$.

Theorem 4.4   *Let $G = (X_i, u_i)_{i \in N} \in \mathbb{G}_q$ and $K \subseteq X$ be compact. If G is concave and diagonally transfer continuous on K, then G is multi-player well-behaved secure on K.*

Proof. The diagonal transfer continuity of G on K implies that, for all $x \in K$, there exists $x'_x \in X$ and $V_x \in N(x)$ such that $U(x'_x, y) > U(y, y)$ for all $y \in V_x$. Since K is compact, there exists a finite subset $\{x_1, \ldots, x_m\}$ of K such that $K \subseteq \cup_{j=1}^m V_{x_j}$ and, by Theorem A.3, a partition of unity $\{\beta_j\}_{j=1}^m$ subordinate to $\{V_{x_j}\}_{j=1}^m$. Define $f : K \to X$ by $f(x) = \sum_{j=1}^m \beta_j(x) x'_{x_j}$. Then, clearly, f is continuous.

We have that, for all $y \in K$, $\beta_j(y) = 0$ if $y \notin V_{x_j}$. Furthermore, if $y \in V_{x_j}$, then $U(x'_{x_j}, y) > U(y, y)$. Since G is concave, then, for all $i \in N$, $u_i(f_i(y), y_{-i}) = u_i(\sum_{j=1}^m \beta_j(y) x'_{x_j, i}, y_{-i}) \geq \sum_{j=1}^m \beta_j(y) u_i(x'_{x_j, i}, y_{-i})$. Hence,

$$U(f(y), y) = \sum_{i=1}^n u_i(f_i(y), y_{-i}) \geq \sum_{i=1}^n \sum_{j=1}^m \beta_j(y) u_i(x'_{x_j, i}, y_{-i})$$

$$= \sum_{j=1}^m \beta_j(y) \sum_{i=1}^n u_i(x'_{x_j, i}, y_{-i}) = \sum_{j=1}^m \beta_j(y) U(x'_{x_j}, y) > U(y, y).$$

Define, for all $x \in K$, $\varphi_x : V_x \rightrightarrows X$ by $\varphi_x(y) = \{f(y)\}$ for all $y \in V_x$. Let $x \in K$ and $y \in V_x$. Then, $U(f(y), y) > U(y, y)$ and, hence, there exists $i \in N$ such that $u_i(f_i(y), y_{-i}) > u_i(y)$. Thus, for all $z \in K$ such that $y \in V_z$, then $\varphi_{z,i}(y) = \{f_i(y)\} \subseteq U_i(y)$ as desired. ∎

From Theorem 4.4 we obtain the existence of Nash equilibrium in games that are concave and diagonally transfer continuous. In fact, if $G \in \mathbb{G}_q$ is concave, diagonally transfer continuous and such that $E(G) = \emptyset$, then G is multi-player well-behaved secure at X by Theorem 4.4. But this, together with $E(G) = \emptyset$, contradicts Theorem 4.1.

Theorem 4.5   *If $G = (X_i, u_i)_{i \in N} \in \mathbb{G}_q$ is concave and diagonally transfer continuous, then G has a Nash equilibrium.*

4.3   Generalized C-Security

We next consider the notion of generalized C-security. As Theorems 4.11 and 4.12 below will show, generalized C-security is clearly related to generalized better-reply security in the following sense. Roughly, generalized better-reply security can be understood as a condition that imposed two inequalities which need to hold with some slack. Generalized C-security requires the same inequalities to hold but without the slack.

The formal definition of generalized C-security is as follows. Let $G = (X_i, u_i)_{i \in N}$ be a normal-form game. For all $\alpha \in \mathbb{R}^n$ and $U \subseteq X$, we say that G is *generalized α-secure on U* if there exists a well-behaved product correspondence $\varphi : U \rightrightarrows X$ such that

(a) $u_i(z_i, y_{-i}) \geq \alpha_i$ for all $i \in N$, $y \in U$ and $z_i \in \varphi_i(y)$, and
(b) For all $y \in U$, there exists $i \in N$ such that $u_i(y) < \alpha_i$.

We say that G is *generalized C-secure on U* if it is generalized α-secure on U for some $\alpha \in \mathbb{R}^n$. A game G is *generalized C-secure at x* if it is generalized C-secure on some open neighborhood U of x. Finally, we say that G is *generalized C-secure* if G is generalized secure at x for all $x \in E(G)^c$.

The relationship between generalized C-security and multi-player well-behaved security is established in the next result. It shows that if a game is generalized C-secure at every point of a compact set, then it is also multi-player well-behaved secure on that compact set.

Theorem 4.6   *Let $G = (X_i, u_i)_{i \in N} \in \mathbb{G}_q$ and K be a compact subset of X. If G is generalized C-secure at x for every $x \in K$ then G is multi-player well-behaved secure on K.*

Theorem 4.6 relies on the following two lemmas. Given $\alpha_1, \ldots, \alpha_m \in \mathbb{R}^n$, let $\max_{j=1,\ldots,m} \alpha_j = (\max_{j=1,\ldots,m} \alpha_1^j, \ldots, \max_{j=1,\ldots,m} \alpha_n^j)$.

Lemma 4.7   *Let $G = (X_i, u_i)_{i \in N} \in \mathbb{G}_q$, $\{\alpha_j\}_{j=1}^m \subseteq \mathbb{R}^n$ and $\{U_j\}_{j=1}^m$ be a collection of open subsets of X such that G is generalized α_j-secure on U_j for all $j \in \{1, \ldots, m\}$. Then G is generalized $\max_{j=1,\ldots,m} \alpha_j$-secure on $\cap_{j=1}^m U_j$.*

Proof. Let, for all $j \in \{1, \ldots, m\}$, $\varphi^j : U_j \rightrightarrows X$ be such that (a) $u_i(z_i, y_{-i}) \geq \alpha_i^j$ for all $i \in N$, $y \in U_j$ and $z_i \in \varphi_i^j(y)$, and (b) for all $y \in U_j$ there exists $i \in N$ such that $u_i(y) < \alpha_i^j$.

Define $\alpha = \max_{j=1,\ldots,m} \alpha_j$ and $U = \cap_{j=1}^m U_j$. Furthermore, for all $i \in N$, let $j(i) \in \{1, \ldots, m\}$ be such that $\alpha_i^{j(i)} = \alpha_i$ and define $\varphi_i = \varphi_i^{j(i)}$.

Then, for all $i \in N$, $y \in U$ and $z_i \in \varphi_i(y)$, we have that $y \in U_{j(i)}$, $z_i \in \varphi_i^{j(i)}$ (y) and, hence, $u_i(z_i, y_{-i}) \geq \alpha_i^{j(i)} = \alpha_i$. Moreover, for all $y \in U$, then $y \in U_1$ and there is $i \in N$ such that $u_i(y) < \alpha_i^1 \leq \alpha_i$. It follows that G is generalized α-secure on U. ∎

Lemma 4.8 *Let $G = (X_i, u_i)_{i \in N} \in \mathbb{G}_q$, K be a compact subset of X and suppose that, for all $x \in K$, there is $\alpha_x \in \mathbb{R}^n$ such that G is generalized α_x-secure at x. Then there exists a function $f = (f_1, \ldots, f_n) : K \to \mathbb{R}^n$ such that G is $f(x)$-secure at x for all $x \in K$ and f_i is upper semi-continuous and finite-valued for all $i \in N$.*

Proof. For all $x \in K$, let $\alpha_x \in \mathbb{R}^n$ and $V_x \in N(x)$ be such that G is generalized α_x-secure on V_x. Since X is a metric space, for all $x \in K$, there exists a closed subset F_x of V_x such that $x \in \text{int}(F_x)$. Indeed, there exists $\varepsilon > 0$ such that the ball $B_\varepsilon(x)$ of radius ε around x is contained in V_x; hence, simply take $F_x = \text{cl}(B_{\varepsilon/2}(x))$.

Since K is compact and $\{\text{int}(F_x)\}_{x \in K}$ is an open cover of K, there exists $\{x_1, \ldots, x_m\} \subseteq K$ such that $K \subseteq \cup_{j=1}^m \text{int}(F_{x_j})$. Define, for all $x \in K$, $J(x) = \{j \in \{1, \ldots, m\} : x \in F_{x_j}\}$ and

$$f(x) = \max\{\alpha_{x_j} : j \in J(x)\}.$$

It is clear that f is finite-valued. To see that f is upper semicontinuous, consider $x \in K$ and a sequence $\{x_k\}_{k=1}^\infty \subseteq K$ converging to x. Letting $j \in \{1, \ldots, m\}$ be such that $\limsup_k f(x_k) = \alpha_{x_j}$, it follows that there exists a subsequence $\{x_{k_l}\}_{l=1}^\infty$ of $\{x_k\}_{k=1}^\infty$ such that $x_{k_l} \in F_{x_j}$ for all $l \in \mathbb{N}$. Hence, $x = \lim_l x_{k_l} \in F_{x_j}$ and, therefore, $f(x) \geq \alpha_{x_j} = \limsup_k f(x_k)$.

Finally, we show that G is generalized $f(x)$-secure for all $x \in K$. Let $x \in K$. Since G is generalized α_{x_j}-secure on V_{x_j} for all $j \in J(x)$ and $f(x) = \max_{j \in J(x)} \alpha_{x_j}$ then G is generalized $f(x)$-secure on $\cap_{j \in J} V_{x_j}$ by Lemma 4.7. Thus, G is generalized $f(x)$-secure at x. ∎

We now turn to the proof of Theorem 4.6.

Proof of Theorem 4.6. Suppose that G is generalized C-secure for every $x \in K$, where $K \subseteq X$ is compact. Then, by Lemma 4.8, there exists a function $f = (f_1, \ldots, f_n) : K \to \mathbb{R}^n$ such that G is $f(x)$-secure at x for all $x \in K$ and f_i is upper semi-continuous and finite-valued for all $i \in N$. Hence, for all $x \in X$, there exists an open neighborhood $\hat{V}_x$ of x such that $f_i(y) \leq f_i(x)$ for all $y \in \hat{V}_x$.

Since G is $f(x)$-secure at x for all $x \in K$, let $\{V_x, \varphi_x\}_{x \in K}$ be as in the definition of generalized C-security and such that $V_x \subseteq \hat{V}_x$ (i.e., starting with $\{\tilde{V}_x, \tilde{\varphi}_x\}_{x \in K}$, define $V_x = \tilde{V}_x \cap \hat{V}_x$ and $\varphi_x = \tilde{\varphi}|_{V_x}$).

For all $x \in K$ and $a \in \mathbb{R}$, let $B_i(x, a) = \{z_i \in X_i : u_i(z_i, x_{-i}) \geq a\}$. Note that for all $x \in X$, $y \in V_x$, $i \in N$ and $z_i \in \varphi_{x,i}(y)$, $u_i(z_i, y_{-i}) \geq f_i(x) \geq f_i(y)$, i.e., $\varphi_{x,i}(y) \subseteq B_i(y, f_i(y))$.

Since K is compact, then, for some finite subset $\{x_1, \ldots, x_m\}$ of K, $K \subseteq \cup_{j=1}^{m} V_{x_j}$. Let $\{\beta_j\}_{j=1}^{m}$ be a partition of unity subordinate to $\{V_{x_j}\}_{j=1}^{m}$ (see Theorem A.3). Define $\Phi : K \rightrightarrows X$ by $\Phi(x) = \sum_{j=1}^{m} \beta_j(x)\varphi_{x_j}(x)$. Then, by Theorem A.8, Φ is well-behaved.

Given the above definition of Φ, we have that, for all $y \in K$, $\beta_j(y) = 0$ if $y \notin V_{x_j}$. Furthermore, if $y \in V_{x_j}$, then, by the above, $\varphi_{x_j,i}(y) \subseteq B_i(y, f_i(y))$ for all $i \in N$. Since $B_i(y, f_i(y))$ is convex (due to $G \in \mathbb{G}_q$), then $\Phi_i(y) \subseteq B_i(y, f_i(y))$ for all $i \in N$.

Define $\hat{\varphi}_x = \Phi|_{V_x}$ for all $x \in K$. Let $x \in K$ and $y \in V_x$. Since $y \in V_y$, then, by part (b) of the definition of generalized C-security, there exists $i \in N$ such that $u_i(y) < f_i(y)$. Furthermore, for all $z \in K$ such that $y \in V_z$, then $\hat{\varphi}_{z,i}(y) = \Phi_i(y) \subseteq B_i(y, f_i(y))$. Hence, for all $w_i \in \hat{\varphi}_{z,i}(y)$, $u_i(w_i, y_{-i}) \geq f_i(y) > u_i(y)$. Thus, $\hat{\varphi}_{z,i}(y) \subseteq U_i(y)$ as desired. ∎

We will show below that $E(G)$ is closed for all $G \in \mathbb{G}_q$ satisfying generalized C-security. Thus, Theorem 4.6 does not imply that every generalized C-secure game $G \in \mathbb{G}$ is multi-player well-behaved secure. Nevertheless, we have the following implication of Theorem 4.6: if $G \in \mathbb{G}$ is generalized C-secure and $E(G) = \emptyset$, then G is multi-player well-behaved secure on X by Theorem 4.6. But this, together with $E(G) = \emptyset$, contradicts Theorem 4.1. Consequently, every generalized C-secure game $G \in \mathbb{G}_q$ has a Nash equilibrium.

Theorem 4.9   *If $G = (X_i, u_i)_{i \in N} \in \mathbb{G}_q$ is generalized C-secure then G has a Nash equilibrium.*

Instead of using Theorem 4.6, we can give a proof of Theorem 4.9 using an argument similar to the one in the proof of Theorem 4.1. Such proof is as follows: Suppose, in order to reach a contradiction, that $E(G) = \emptyset$. This implies that $E(G)^c = X$ and that G is generalized C-secure at X. Hence, by Lemma 4.8, there exists a function $f = (f_1, \ldots, f_n) : X \to \mathbb{R}^n$ such that G is $f(x)$-secure at x for all $x \in X$ and f_i is upper semi-continuous and finite-valued for all $i \in N$.

We next consider the game $\underline{G} = (X_i, \underline{u}_i)_{i \in N}$, where $\underline{u}$ is as in Chapter 3. Let $E(\underline{G}, f)$ be the set of f-equilibria of $\underline{G}$ and $B_{(\underline{G},f)} : X \rightrightarrows X$ be the

f-best-reply correspondence of $\underline{G}$ defined by

$$B_{(\underline{G},f)}(x) = \{y \in X : \underline{u}_i(y_i, x_{-i}) \geq f(x) \text{ for all } i \in N\}$$

for all $x \in X$. The key properties of the game $\underline{G}$ are: (1) $E(\underline{G}, f) = \emptyset$ and (2) there is a well-behaved correspondence $\Psi : X \rightrightarrows X$ such that $\Psi(x) \subseteq B_{(\underline{G},f)}(x)$ for all $x \in X$ (see Carmona (2011a) for a proof). But (1) and (2) together contradict Cauty's fixed point theorem (Theorem A.14).

Generalized C-security also implies that the set of Nash equilibria is closed.

Theorem 4.10 *If $G = (X_i, u_i)_{i \in N} \in \mathbb{G}$ is generalized C-secure then $E(G)$ is closed.*

Proof. Let $x \in E(G)^c$ and let U_x, φ_x and α_x be as in the definition of generalized C-security. Then, for all $y \in U_x$, there exists $i \in N$ such that $u_i(y) < \alpha_{x,i}$. Since, for all $z_i \in \varphi_{x,i}(y)$, we have that $u_i(z_i, y_{-i}) \geq \alpha_{x,i}$, then $u_i(z_i, y_{-i}) > u_i(y)$. Hence, $y \in E(G)^c$ and $V_x \subseteq E(G)^c$. It follows that $E(G)^c$ is open and that $E(G)$ is closed. ∎

The importance of generalized C-security is partly due to the fact that it extends the notion of generalized better-reply security. The relationship between these two concepts will be clearly seen through a further characterization of generalized better-reply security provided in what follows.

Let $G = (X_i, u_i)_{i \in N} \in \mathbb{G}$ be given. For all $x \in X$, let $A(x) = \{\alpha \in \mathbb{R}^n : (x, \alpha) \in \text{cl}(\text{graph}(u))\}$. We say that G is *generalized better-reply secure* at $x \in X$ if, for all $\alpha \in A(x)$, there is $i \in N$, an open neighborhood U of x_{-i}^*, a well-behaved correspondence $\varphi_i : U \rightrightarrows X_i$ and $\varepsilon > 0$ such that $u_i(x') \geq \alpha_i + \varepsilon$ for all $x' \in \text{graph}(\varphi_i)$. Note that, similarly to Theorem 3.3, G is generalized better-reply secure at $x \in X$ if and only if for all $\alpha \in A(x)$ there exists $i \in N$ such that $\underline{v}_i(x_{-i}) > \alpha_i$.

Generalized better-reply security is equivalent to the following stronger version of generalized C-security. A game $G = (X_i, u_i)_{i \in N} \in \mathbb{G}$ is *B-secure* at $x \in X$ if there is $\alpha \in \mathbb{R}^n$, $\varepsilon > 0$, an open neighborhood U of x and a well-behaved product correspondence $\varphi : U \rightrightarrows X$ such that

(i) $u_i(z_i, y_{-i}) \geq \alpha_i + \varepsilon$ for all $i \in N$, $y \in U$ and $z_i \in \varphi_i(y)$, and
(ii) For all $y \in U$, there exists $i \in N$ such that $u_i(y) < \alpha_i - \varepsilon$.

Theorem 4.11 *Let $G = (X_i, u_i)_{i \in N} \in \mathbb{G}$ and $x \in X$. Then, G is generalized better-reply secure at x if and only if G is B-secure at x.*

Proof. (Sufficiency) Suppose that G is B-secure at x and let α, ε, U and φ be as in its definition. Let $\alpha' \in A(x)$. Then there exists a sequence $\{x_k\}_{k=1}^{\infty} \subseteq X$ such that $\lim_k(x_k, u(x_k)) = (x, \alpha')$. Since N is finite, it follows by (ii) that there is $i \in N$ such that $\alpha_i' \leq \alpha_i - \varepsilon$. Then (i) implies that $u_i(z_i, y_{-i}) \geq \alpha_i + \varepsilon \geq \alpha_i' + 2\varepsilon$ for all $y \in U$ and $z_i \in \varphi_i(y)$. Thus, G is generalized better-reply secure at x.

(Necessity) Suppose that G is generalized better-reply secure at x. Thus, for all $\alpha' \in A(x)$ there exists $i \in N$ such that $\underline{v}_i(x_{-i}) > \alpha_i'$. Hence, for all $\alpha' \in A(x)$, there exists $i_{\alpha'} \in N$, $\varepsilon_{\alpha'} > 0$ and $U_{\alpha'} \in N(\alpha')$ such that $\underline{v}_{i_{\alpha'}}(x_{-i}) > \tilde{\alpha}_{i_{\alpha'}} + \varepsilon_{\alpha'}$ for all $\tilde{\alpha} \in U_{\alpha'}$.

Since $A(x)$ is compact, there exists $\{\alpha_1', \ldots, \alpha_m'\}$ such that $A(x) \subseteq \cup_{j=1}^{m} U_{\alpha_j'}$. Letting $\varepsilon = \min_j \varepsilon_{\alpha_j'}/3$, it follows that, for all $\alpha' \in A(x)$, $\alpha' \in U_{\alpha_j'}$ for some j, and, hence, $\underline{v}_{i_{\alpha_j'}}(x_{-i}) > \alpha_{i_{\alpha_j'}}' + \varepsilon_{\alpha_j'} \geq \alpha_{i_{\alpha_j'}}' + 3\varepsilon$. In conclusion,

$$\text{for all } \alpha' \in A(x), \text{there is } i \in N \text{such that } \underline{v}_i(x_{-i}) > \alpha_i' + 3\varepsilon. \tag{4.1}$$

Define $\alpha \in \mathbb{R}^n$ by $\alpha_i = \underline{v}_i(x_{-i}) - 2\varepsilon$ for all $i \in N$. Hence, $\alpha_i + \varepsilon < \underline{v}_i(x_{-i})$ for all $i \in N$ and, therefore, there is an open neighborhood U of x and a well-behaved product correspondence $\varphi : U \rightrightarrows X$ such that (i) holds.

Suppose, in order to reach a contradiction, that (ii) is false. Hence, we obtain a sequence $\{y_k\}_{k=1}^{\infty} \subseteq X$ such that $\lim_k y_k = x$ and $u_i(y_k) \geq \alpha_i - \varepsilon$ for all $i \in N$ and $k \in \mathbb{N}$. Since u is bounded, we may assume that $\{u(y_k)\}_{k=1}^{\infty}$ converges. Letting $\alpha' = \lim_k u(y_k)$, it follows that $\alpha' \in A(x)$ and $\alpha_i' \geq \alpha_i - \varepsilon = \underline{v}_i(x_{-i}) - 3\varepsilon$ for all $i \in N$, contradicting (4.1).

Hence, there is an open neighborhood V of x such that (ii) holds. Finally, note that (i) holds when U is replaced with $U \cap V$ and φ with $\varphi|_{U \cap V}$. Thus, G is B-secure at x. ∎

It is clear from Theorem 4.11 that all generalized better-reply secure games $G \in \mathbb{G}$ are generalized C-secure.

Theorem 4.12   *If $G = (X_i, u_i)_{i \in N} \in \mathbb{G}$ is generalized better-reply secure then G is generalized C-secure.*

The extra generality of generalized C-security as compared to generalized better-reply security is achieved by dropping the ε in the definition of B-security. This is illustrated by the following example, which shows that generalized C-security is strictly weaker than generalized better-reply security.

Example 4.13 Let $G = (X_i, u_i)_{i \in N}$ be defined as follows: $N = \{1, 2\}$, $X_1 = X_2 = [0, 1]$,

$$
u_1(x_1, x_2) = \begin{cases}
1 & \text{if } x_1 = 1, \\
0 & \text{if } x_1 < 1 \text{ and } x_2 = 1/2, \\
2x_2 & \text{if } x_1 < 1 \text{ and } x_2 < 1/2, \\
2(1 - x_2) & \text{if } x_1 < 1 \text{ and } x_2 > 1/2,
\end{cases}
$$

and $u_2(x_1, x_2) = u_1(x_2, x_1)$ for all $(x_1, x_2) \in X$.

We have that, for all $i \in N$ and $x_{-i} \in X_{-i}$, the set $\{x_i \in X_i : u_i(x_i, x_{-i}) \geq \alpha\}$ is either $\emptyset$, $\{1\}$ or $[0, 1]$. Thus, $u_i(\cdot, x_{-i})$ is quasi-concave.

We have that $B(x) = \{(1, 1)\}$ for all $x \in X$. Hence, $N(G) = \{(1, 1)\}$. Furthermore, it is easy to see that $\underline{u} = u$.

To see that G is not generalized better-reply secure, consider $x^* = (1/2, 1/2)$ and $u^* = (1, 1)$. Then $(x^*, u^*) \in \text{cl}(\text{graph}(u))$ (for instance, letting $x_k = (1/2 - 1/(2k), 1/2 - 1/(2k))$ for all $k \in \mathbb{N}$, we have that $\lim_k (x_k, u(x_k)) = (x^*, u^*))$ and x^* is not a Nash equilibrium of G. But, for all $i \in N$, $u_i^* = 1 \geq v_i(x_{-i}^*) = \underline{v}_i(x_{-i}^*)$.

Finally, we show that G is generalized C-secure. Let $x^* \notin N(G)$ and define $\alpha = (1, 1)$, $U = \{(1, 1)\}^c$ and $\varphi(x) = \{(1, 1)\}$ for all $x \in U$. Then, for all $x \in U$, $u_i(y_i, x_{-i}) \geq \alpha_i$ for all $y_i \in \varphi_i(x)$ (since $y_i = 1$ for all $i \in N$) and there is $i \in N$ such that $u_i(x) < \alpha_i$ (since, for some i, $x_i \neq 1$).

4.4 Lower Single-Deviation Property

We consider in this section the lower single-deviation property. This notion is related to better-reply security in the sense that it uses the same regularization of players' payoff functions (namely, $\bar{u}$ in a game $G = (X_i, u_i)_{i \in N}$). However, better-reply security and the lower single-deviation property differ in the way u and $\bar{u}$ are related.

Let $G = (X_i, u_i)_{i \in N}$ be a normal-form game and recall the definition of $\bar{u}$: For all $i \in N$ and $x \in X$,

$$
\bar{u}_i(x) = \sup_{U \in N(x_{-i})} \inf_{y_{-i} \in U} u_i(x_i, y_{-i}).
$$

For all $Y \subseteq X$, we say that G has the *lower single-deviation property on* Y if, whenever $x^* \in Y$, there exists $\hat{x} \in X$ and an open neighborhood U of x^* such that, for all $x' \in U$, there is a player $i \in N$ for whom $\bar{u}_i(\hat{x}_i, y_{-i}) > \bar{u}_i(x')$ for all $y \in U$. Furthermore, $G \in \mathbb{G}$ has the *lower* *single-deviation property* if it has the lower single-deviation property on $E(G)^c$.

The lower single-deviation property is sufficient for the existence of Nash equilibria.

Theorem 4.14　　*If $G = (X_i, u_i)_{i \in N} \in \mathbb{G}_q$ has the lower single-deviation property then G has a Nash equilibrium.*

The proof of Theorem 4.14 follows the same lines as the previous ones. Suppose that $E(G) = \emptyset$, which implies that $E(G)^c = X$ and that G has the lower single-deviation property on X. We then consider the game $\bar{G} = (X_i, \bar{u}_i)_{i \in N} \in \mathbb{G}$ and show that (1) $E(\bar{G}) = \emptyset$ and (2) $\bar{G}$ is generalized C-secure. But (1) and (2) together contradict Theorem 4.9.

Proof of Theorem 4.14.　　Suppose that G has the lower single-deviation property and $E(G) = \emptyset$. Thus, for all $x \in X$, there exists $\hat{x} \in X$ and an open neighborhood U of x such that, for all $x' \in U$, there exists $i \in N$ such that $\bar{u}_i(\hat{x}_i, y_{-i}) > \bar{u}_i(x')$ for all $y \in U$. In particular, when $x' = y = x$, it follows that there exists $i \in N$ such that $\bar{u}_i(\hat{x}_i, x_{-i}) > \bar{u}_i(x)$. Hence, $x \notin E(\bar{G})$ for all $x \in X$, i.e. $E(\bar{G}) = \emptyset$.

We next show that $\bar{G}$ is generalized C-secure. Let $x \in X$ and consider U and $\hat{x}$ as above. Since X is a metric space, then there is a closed set $F \subseteq U$ such that $x \in \text{int}(F)$. Define

$$\alpha_x = \left(\min_{y \in F} \bar{u}_1(\hat{x}_1, y_{-1}), \ldots, \min_{y \in F} \bar{u}_n(\hat{x}_n, y_{-n}) \right).$$

Since F is compact and $\bar{u}_i(\hat{x}_i, \cdot)$ is lower semi-continuous for all $i \in N$ (by Lemma 3.11), α_x is well-defined.

We claim that $\bar{G}$ is $(\alpha_x, \text{int}(F))$-secure at x. To establish this claim, define $\varphi : \text{int}(F) \rightrightarrows X$ by $\varphi(z) = \{\hat{x}\}$ for all $z \in \text{int}(F)$. Thus, for all $i \in N$ and $z \in \text{int}(F)$, $\bar{u}_i(\hat{x}_i, z_{-i}) \geq \min_{y \in F} \bar{u}_i(\hat{x}_i, y_{-i}) = \alpha_{x,i}$. Furthermore, since $\text{int}(F) \subseteq F \subseteq U$, for all $z \in \text{int}(F)$, there exists $i \in N$ such that $\bar{u}_i(\hat{x}_i, y_{-i}) > \bar{u}_i(z)$ for all $y \in F$. Hence, $\alpha_{x,i} = \min_{y \in F} \bar{u}_i(\hat{x}_i, y_{-i}) > \bar{u}_i(z)$.

Since $\bar{G}$ is $(\alpha_x, \text{int}(F))$-secure at x for all $x \in X$ and $E(G) = \emptyset$, it follows that $\bar{G}$ is generalized C-secure. But this, together with $E(G) = \emptyset$, contradicts Theorem 4.9. ∎

The use of $\bar{u}$ in the definition of lower single-deviation property suggests that this notion is related with better-reply security and with its generalization achieved through reduced security relative to $\bar{u}$. Theorem 4.15 below shows that the lower single-deviation property is even weaker than reduced security relative to $\bar{u}$. Furthermore, it shows that the two concepts coincide in *finite-valued* games, i.e. when $G = (X_i, u_i)_{i \in N}$ is such that $u_i(X)$ is finite for all $i \in N$. The latter result is due to the fact that the extra generality of the lower single-deviation property as compared to reduced security relative to $\bar{u}$ is achieved by dropping the ε in the definition of reduced security relative to $\bar{u}$.

Theorem 4.15 *Let* $G = (X_i, u_i)_{i \in N} \in \mathbb{G}$. *Then the following holds*:

(a) *If G is reducible secure relative to $\bar{u}$, then G has the lower single-deviation property.*
(b) *If G is finite-valued and has the lower single-deviation property, then G is reducible secure relative to $\bar{u}$.*

Proof. Suppose that G is reducible secure relative to $\bar{u}$ and let $x \in E(G)^c$. Let $\varepsilon > 0$ and U be such that, for all $x' \in U$, there is $i \in N$ for whom $\bar{v}_i(x_{-i}) \geq \bar{u}_i(x') + \varepsilon$. For all $i \in N$, let $\hat{x}_i \in X_i$ be such that $\bar{u}_i(\hat{x}_i, x_{-i}) > \bar{v}_i(x_{-i}) - \varepsilon/2$. Furthermore, for all $i \in N$, $\bar{u}_i(\hat{x}_i, \cdot)$ is lower semi-continuous and, hence, there exists an open neighborhood V of x such that $\bar{u}_i(\hat{x}_i, y_{-i}) > \bar{u}_i(\hat{x}_i, x_{-i}) - \varepsilon/2$ for all $i \in N$.

Note that $V \cap U$ is an open neighborhood of x and let $\hat{x} = (\hat{x}_1, \ldots, \hat{x}_n)$. Let $x' \in V \cap U$. Then, there is $i \in N$ such that $\bar{v}_i(x_{-i}) \geq \bar{u}_i(x') + \varepsilon$. Hence, for all $y \in V$,

$$\bar{u}_i(\hat{x}_i, y_{-i}) > \bar{u}_i(\hat{x}_i, x_{-i}) - \varepsilon/2 > \bar{v}_i(x_{-i}) - \varepsilon \geq \bar{u}_i(x').$$

Thus, G has the lower single-deviation property.

We next establish (b). Suppose that G is finite-valued and has the lower single-deviation property. Since G is finite-valued, then $\bar{u}_i(X)$ is finite. Thus, there exists $\varepsilon > 0$ such that $\bar{u}_i(x) > \bar{u}_i(y)$ implies that $\bar{u}_i(x) > \bar{u}_i(y) + \varepsilon$. Let $x \in E(G)^c$ and let $\hat{x} \in X$ and U be an open neighborhood of x has in the definition of the lower single-deviation property. Then, for all $x' \in U$ there exists $i \in N$ such that $\bar{u}_i(\hat{x}, x_{-i}) > \bar{u}_i(x')$. Hence, $\bar{v}_i(x_{-i}) \geq \bar{u}_i(\hat{x}, x_{-i}) > \bar{u}_i(x') + \varepsilon$. ∎

We next present two examples showing that the lower single-deviation property is, in general, related neither to multi-player well-behaved security nor to generalized C-security.

The first example is an example of a game $G \in \mathbb{G}_q$ that has the lower single-deviation property but is neither multi-player well-behaved secure nor generalized C-secure. Recall Example 3.16: G is such that $N = \{1, 2\}$,

$X_1 = X_2 = [0, 1]$,

$$u_1(x) = \begin{cases} 2 & \text{if } x_1 \geq 3/4, \\ 2 & \text{if } x_2 = 1 \text{ and } x_1 > 1/2, \\ 1 & \text{if } x_2 < 1 \text{ and } 1/2 < x_1 < 3/4, \\ 0 & \text{otherwise,} \end{cases}$$

and

$$u_2(x) = \begin{cases} 2 & \text{if } x_2 = 1, \\ 0 & \text{otherwise} \end{cases}$$

for all $x \in X$.

We have shown that G is reducible secure relative to $\bar{u}$. Thus, since G is finite-valued, G has the lower single-deviation property. Furthermore, we have that $E(G) = (1/2, 1] \times \{1\}$ and, hence, $E(G)$ is not closed. This, together with Theorems 4.3 and 4.10, implies that G is neither multi-player well-behaved secure nor generalized C-secure.

The second example is an example of a game $G \in \mathbb{G}_q$ which is multi-player well-behaved secure but does not have the lower single-player deviation. Furthermore, we will show that such game is also not generalized C-secure.

Example 4.16   Let G be such that $N = \{1, 2\}$, $X_1 = X_2 = [0, 1]$,

$$u_1(x) = \begin{cases} 2 & \text{if } x_2 = x_1/2 \text{ and } x_1 < 1, \\ 1 & \text{if } x_2 < x_1/2 \text{ or } x_1 = 1, \\ 0 & \text{otherwise,} \end{cases}$$

and

$$u_2(x) = \begin{cases} 2 & \text{if } x_2 = 1 \text{ and } x_1 < 1, \\ 0 & \text{if } x_2 < 1 \text{ and } x_1 = 1, \\ 1 & \text{otherwise.} \end{cases}$$

for all $x \in X$.

It is clear that $E(G) = \{(1, 1)\}$. Furthermore, we have that

$$\bar{u}_1(x) = \begin{cases} 1 & \text{if } x_2 < x_1/2 \text{ or } x_1 = 1, \\ 0 & \text{otherwise,} \end{cases}$$

$\bar{u}_2 = u_2$, $\bar{v}_1 \equiv 1$ and

$$\bar{v}_2(x_1) = \begin{cases} 2 & \text{if } x_1 < 1, \\ 1 & \text{if } x_1 = 1. \end{cases}$$

Let $x = (1, 1/2) \in E(G)^c$, $\varepsilon > 0$ and U be an open neighborhood of x. Then, there exists $x' \in U$ such that $x'_1 < 1$ and $x'_2 < x'_1/2$. Hence, for all $i \in N$, $\bar{v}_i(x_{-i}) = 1 < 1 + \varepsilon = \bar{u}_i(x') + \varepsilon$. This shows that G is not reducible secure relative to $\bar{u}$. Since G is finite-valued, then G does not have the lower single-deviation property.

We also note that G is not generalized C-secure. Consider again $x = (1, 1/2) \in E(G)^c$ and suppose that there exist U, φ and α as in the definition of generalized C-security. Since there exists $y \in U$ such that $x_1 < 1$ and $x_2 = x_1/2$, therefore, satisfying $u(y) = (2, 1)$, it follows by part (b) of the definition of generalized C-security that either $\alpha_1 > 2$ or $\alpha_2 > 1$. Since $u_1(z_1, x_2) = 1$ for all $z_1 \in X_1$, then part (a) of the definition of generalized C-security implies that $\alpha_1 \leq 1$. But since $u_2(x_1, z_2) \leq 1$ for all $z_2 \in X_2$, then (a) also implies that $\alpha_2 \leq 1$, contradicting the above conclusion. Hence, G is not generalized C-secure.

We next show that G is multi-player well-behaved secure. For all $x \in E(G)^c$, let $V_x = \{(1, 1)\}^c$ and $\varphi_x(y) = \{(1, 1)\}$ for all $y \in V_x$. Let $x \in E(G)^c$ and $y \in V_x$. If $y_2 = 1$, then $y_1 < 1$ and $\varphi_{z,1}(y) = \{1\} \in U_1(y)$ for all $z \in E(G)^c$. Otherwise, then $y_2 < 1$ and $\varphi_{z,2}(y) = \{1\} \in U_2(y)$ for all $z \in E(G)^c$. Thus, G is multi-player well-behaved secure.

4.5 Generalized Weak Transfer Continuity

The last concept from the recent literature on existence of equilibrium that we consider is that of generalized weak transfer continuity. This notion is related to generalized better-reply security as follows. Recall that in generalized better-reply secure games, whenever a strategy is not a Nash equilibrium, some player can secure a payoff strict above a given limit payoff. Also, note that non-Nash equilibrium strategies are such that some player can gain by deviating from such strategy. Roughly, generalized weak transfer continuity requires that, whenever a strategy is not a Nash equilibrium, some player can secure a strict positive deviation gain.

For all $Y \subseteq X$, we say that a normal-form game $G = (X_i, u_i)_{i \in N}$ is *generalized weakly transfer continuous on Y* if whenever $x \in Y$, there exists a player $i \in N$, an open neighborhood U of x and a well-behaved correspondence $\varphi_i : U \rightrightarrows X_i$ such that $\inf_{(y,z_i) \in \text{graph}(\varphi_i)} (u_i(z_i, y_{-i}) - u_i(y)) > 0$. A normal-form game $G \in \mathbb{G}$ is *generalized weakly transfer continuous* if it is generalized weakly transfer continuous on $E(G)^c$.

The next result shows that every generalized weakly transfer continuous game $G \in \mathbb{G}_q$ has a Nash equilibrium.

Theorem 4.17   *If $G = (X_i, u_i)_{i \in N} \in \mathbb{G}_q$ is generalized weakly transfer continuous then G has a Nash equilibrium.*

The proof follows the same lines as the previous ones. Suppose that $E(G) = \emptyset$, which implies that $E(G)^c = X$ and that G is generalized weakly transfer continuous at X.

We consider the game $G_0 = (Z_i, r_i)_{i \in N_0}$ defined as follows: $N_0 = N \cup \{0\}$, $Z_0 = X_1 \times \cdots \times X_n$, $Z_i = X_i$ and, for all $z \in Z$,

$$r_i(z) = u_i(z_i, z_{0,-i}) - u_i(z_0)$$

for all $i \in N$ and

$$r_0(z) = \begin{cases} 1 & \text{if } z_0 = z_{-0}, \\ 0 & \text{otherwise.} \end{cases}$$

In the above formula for r_i with $i \in N$, note that $z_0 = (z_{0,1}, \ldots, z_{0,n})$ and, therefore, $z_{0,-i} = (z_{0,1}, \ldots, z_{0,i-1}, z_{0,i+1}, \ldots, z_{0,n})$. Letting $z_{-0,i}$ denote z without the coordinates of players 0 and i, we have that, if $z_0 = z_{-0}$ then $z_{0,-i} = z_{-0,i}$. We next show that (1) $E(G_0) = \emptyset$ and (2) G_0 is generalized better-reply secure. But (1) and (2) together contradict Theorem 3.2.

Proof of Theorem 4.17.   We first show that $E(G_0) = \emptyset$. Suppose that $E(G_0) \neq \emptyset$ and let $z^* \in E(G_0)$. For convenience, let $x^* = z^*_{-0}$. Then $r_0(z^*) \geq r_0(z_0, z^*_{-0})$ for all $z_0 \in Z_0$ implies $z^*_0 = z^*_{-0} = x^*$. Furthermore, for all $i \in N$, since $r_i(z^*) = u_i(z^*_i, z^*_{0,-i}) - u_i(z^*_0) = 0$ (due to $z^*_0 = z^*_{-0}$) and, for all $x_i \in Z_i = X_i$, $r_i(x_i, z^*_{-i}) = u_i(x_i, x^*_{-i}) - u_i(x^*)$ (again due to $z^*_0 = x^*$), then $r_i(z^*) \geq r_i(x_i, z^*_{-i})$ for all $i \in N$ and $x_i \in X_i$ implies that $u_i(x^*) \geq u_i(x_i, x^*_{-i})$ for all $i \in N$ and $x_i \in X_i$. Hence, $x^* \in E(G)$, a contradiction.

We next show that G_0 is generalized better-reply secure. Let $(z^*, u^*) \in \text{cl}(\text{graph}(r))$. We consider two cases. The first case is when $u^*_0 \neq 1$. In this case, the definition of r_0 implies that $u^*_0 = 0$. Hence, define $V_{z^*_{-0}} = Z_{-0}$ and $\varphi_0 : V_{z^*_{-0}} \rightrightarrows Z_0$ by $\varphi_i(s) = s$ for all $s \in V_{z^*_{-0}}$ and note that, for all $z \in \text{graph}(\varphi_0)$, $r_0(z) = 1 > 0 = u^*_0$. This shows that G_0 is generalized better-reply secure in the case where $u^*_0 \neq 1$.

The second case is when $u_0^* = 1$. Let $\{z_k\}_{k=1}^\infty$ be a sequence converging to z^* such that $\lim_k r(z_k) = u^*$. Then, there exists $K \in \mathbb{N}$ such that $z_0^k = z_{-0}^k$ for all $k \geq K$. Hence, for all $i \in N$ and $k \geq K$, $r_i(z_k) = u_i(z_i^k, z_{0,-i}^k) - u_i(z_0) = 0$, which implies that $u_i^* = 0$. Note also that $z_0^k = z_{-0}^k$ for all $k \geq K$ implies that $z_0^* = z_{-0}^*$.

Since G is generalized weakly transfer continuous at X, then there exists $i \in N$, an open neighborhood V of $z_{-0}^* = z_0^*$, a well-behaved correspondence $\psi_i : V \rightrightarrows Z_i$ and $\varepsilon > 0$ such that $u_i(y_i, x_{-i}) - u_i(x) \geq \varepsilon$ for all $(x, y_i) \in \text{graph}(\psi_i)$. Define $U = V \times X$ and $\varphi_i : U \rightrightarrows Z_i$ by $\varphi_i(x, s) = \psi_i(x)$ for all $(x, s) \in V \times X$. Let $(x, s) \in V \times X$ and $y_i \in \varphi_i(x, s)$ and, for convenience, define $z_0 = x$ and $z_{-0} = (y_i, s_{-i})$. Then, $z_0 \in V$, $y_i \in \psi_i(z_0)$ and so $r_i(z_0, z_{-0}) = u_i(y_i, x_{-i}) - u_i(x) \geq \varepsilon > 0 = u_i^*$. This shows that G_0 is generalized better-reply secure in the case where $u_0^* = 1$ and completes the proof. ∎

4.6 References

The notion of multi-player well-behaved security is due to Barelli and Soza (2010), as well as Theorem 4.1 and Lemma 4.2.

Diagonal transfer continuity was introduced by Baye, Tian and Zhou (1993) and Theorem 4.4 is based on Proposition 3.7 in Barelli and Soza (2010).

McLennan, Monteiro and Tourky (2011) introduced the notion of C-security, while its generalized version, generalized C-security, is due to Barelli and Soza (2010). Lemmas 4.7 and 4.8 were first established by McLennan, Monteiro and Tourky (2011) and then generalized by Barelli and Soza (2010), who also established a result analogous to Theorem 4.6. Theorems 4.11 and 4.12 are due to McLennan, Monteiro and Tourky (2011).

The notion of lower single-player security and Theorem 4.14 are due to Reny (2009). Theorem 4.15 and Example 4.16 are due to Carmona (2011a). Generalized weak transfer continuity was introduced by Nessah (2011), who established Theorem 4.17.

The proof of Theorems 4.1, 4.14 and 4.17 are based on Carmona (2011a).

Several of the conditions considered in this chapter also allow for non-quasiconcave games in its original formalization. This is the case of multi-player well-behaved security and generalized C-security; see Barelli and

Soza (2010) and McLennan, Monteiro and Tourky (2011). Note also that Baye, Tian and Zhou (1993), who introduced diagonal transfer continuity, also allow for non-quasiconcave games. See also Bich (2009) for an extension of better-reply security to the case of non-quasiconcave games.

Barelli and Soza (2010) also allow for qualitative games, i.e., games where players' preferences are described by preference relations. See also Reny (2011a) for further results on qualitative games.

Chapter 5

Limit Results

Limit results, such as the ones in Section 2.4 of Chapter 2, provide a useful tool to address the question of existence of equilibrium. This was illustrated by the existence result for generalized better-reply secure games, which relied partly on a limit argument.

Furthermore, limit results are also important because they are a way of providing a sense in which equilibria are robust. The notion of robustness that we consider concerns how the set of Nash equilibria of a game changes with changes in players' payoff functions and strategy spaces.

We start by presenting a notion of convergence of payoff functions and strategy spaces, therefore, of games, and a limit result for better-reply closed games (relative to themselves). This limit result concludes that limit points of approximate equilibria (in the sense of f-equilibria) of a sequence of games converging to a limit, better-reply closed, game are themselves Nash equilibria of the limit game.

The above result requires the sequence $\{f_k\}_{k=1}^{\infty}$ of target functions to approximate players' value functions in the limit game in a given sense. Since this requirement is met by considering sequences of ε-equilibrium with ε converging to zero, we are led to consider limit result for this notion of approximate equilibria.

We then present the notion of sequential better-reply security of Carbonell-Nicolau and McLean (2011) and two forms of limit results for ε-equilibria. We show that both limit results are equivalent to the sequential better-reply security of the limit game. We then consider several variations of sequential better-reply security as well as sufficient conditions for it.

Given the focus on ε-equilibria, we present two existence result for this equilibrium concept that are not covered by the results presented previously.

The limit result in this chapter are then used to establish the upper hemicontinuity of the equilibrium correspondence on the set of sequentially better-reply secure games and its generic continuity. The interpretation of this result is that, except in exceptional cases, the Nash equilibria of a large class of discontinuous games are robust to changes in players' payoff functions and strategy spaces.

We conclude this chapter with an alternative notion of robustness of Nash equilibria of infinite action spaces due to Reny (2011b).

5.1   A General Limit Result

The setting needed to formulate limit results is analogous to the one considered in Section 2.4 of Chapter 2, the main difference being that payoff functions are not assumed to be continuous.

Let N be a finite set of players and $(X_i)_{i \in N}$ be a collection of compact subsets of a metric space. The space of games consists of all compact metric games $G = (Y_i, u_i|_{Y_i})_{i \in N}$ with $Y_i \subseteq X_i$ and $u_i : X \to \mathbb{R}$ for all $i \in N$ (i.e., even though the strategy spaces are $(Y_i)_{i \in N}$, each player's payoff function is defined on X). We denote this space by $\mathbb{G}(X)$. When $G = (Y_i, u_i|_{Y_i})_{i \in N}$ is such that, in addition, Y_i is a convex subset of a vector space and $u_i(\cdot, x_{-i})$ is quasiconcave for all $i \in N$ and $x_{-i} \in X_{-i}$, we write $G \in \mathbb{G}_q(X)$. To simplify the notation, we will write $G = (Y_i, u_i)_{i \in N}$ instead of $G = (Y_i, u_i|_{Y_i})_{i \in N}$. Furthermore, the subset of $\mathbb{G}(X)$ consisting of games $G = (Y_i, u_i)_{i \in N}$ with $Y_i = X_i$ for all $i \in N$ is denoted by $\mathbb{G}_s(X)$.

We say that two games are close to each other if both players' payoff functions and players' strategy spaces are closed to each other. The notion of distance between payoff functions that we use is the sup norm: Given $G = (Y_i, u_i)_{i \in N}$ and $G' = (Y_i, u_i')_{i \in N}$, let $\|u - u'\| = \max_{i \in N} \sup_{x \in X} |u_i(x) - u_i'(x)|$. The notion of distance between strategy sets we use is the Hausdorff distance, defined as follows. Given a metric space Z with metric d, let, for all $z \in Z$ and $A \subseteq Z$, $d(z, A) = \inf\{d(z, y) : y \in A\}$. Then, letting d_i be a metric on X_i for all $i \in N$, define $\delta_i(Y_i, Y_i') = \max \{\sup_{x_i \in Y_i} d_i(x_i, Y_i'), \sup_{x_i' \in Y_i'} d_i(x_i', Y_i)\}$ and $\delta(Y, Y') = \max_{i \in N} \delta_i(Y_i, Y_i')$.

The assumption that players' strategy spaces are compact and, therefore, closed subsets of X_i for all $i \in N$, allows for the following characterization of convergent sequences of strategy spaces. First, we introduce some definition for a general metric space Z. Given a sequence $\{E_k\}_{k=1}^{\infty} \subseteq Z$, the *topological limsup* of $\{E_k\}_{k=1}^{\infty}$, denoted $\mathrm{Ls}(E_k)$, consists of the points $z \in Z$

such that, for every neighborhood V of z, there exist infinitely many $k \in \mathbb{N}$ with $V \cap E_k \neq \emptyset$. Moreover, the *topological liminf* of $\{E_k\}_{k=1}^{\infty}$, denoted $\mathrm{Li}(E_k)$, consists of the points $z \in Z$ such that, for every neighborhood V of z, we have $V \cap E_k \neq \emptyset$ for all but finitely many $k \in \mathbb{N}$. In the particular case considered above, given $i \in N$ and $\{Y_i^k\}_{k=1}^{\infty} \subseteq X_i$, we obtain that $\lim_k \delta_i(Y_i^k, Y_i) = 0$ if and only if $Y_i = \mathrm{Ls}(Y_i^k) = \mathrm{Li}(Y_i^k)$ (see Theorem A.2).

Finally, we make $\mathbb{G}(X)$ a metric space by defining $d(G, G') = \max\{\|u - u'\|, \delta(Y, Y')\}$ for all $G, G' \in \mathbb{G}(X)$. We have that a sequence $\{G_k\}_{k=1}^{\infty} \subseteq \mathbb{G}(X)$ converges to G if and only if $\{u_i^k\}_{k=1}^{\infty}$ converges to u_i uniformly and $Y_i = \mathrm{Ls}(Y_i^k) = \mathrm{Li}(Y_i^k)$ for all $i \in N$.

The following result shows that better-reply closedness of a game G relative to its payoff function is sufficient for limit points of sequences of approximate equilibria of games converging to G to be Nash equilibria of G.

Theorem 5.1   Let $G \in \mathbb{G}$ be better-reply closed relative to u, $\{G_k\}_{k=1}^{\infty} \subseteq \mathbb{G}(X)$, $\{f_k\}_{k=1}^{\infty}$ be such that $f_k \in F(G_k)$ for all $k \in \mathbb{N}$, $\{x_k\}_{k=1}^{\infty} \subseteq X$ and $x \in X$.

If $G = \lim_k G_k$, $x = \lim_k x_k$, x_k is an f_k-equilibrium of G_k for all $k \in \mathbb{N}$ and $\liminf_k f_i^k(x_k) \geq v_i(x_{-i})$ for all $i \in N$, then x is Nash equilibrium of G.

Proof.   Since u is bounded, we may assume that $\{u(x_k)\}_{k=1}^{\infty}$ converges; let $u^* = \lim_k u(x_k)$. Since $\lim_k u_k = u$, it follows that $\lim_k u_k(x_k) = u^*$. Then $u_i^* = \lim_k u_i^k(x_k) \geq \liminf_k f_i^k(x_k) \geq v_i(x_{-i})$ for all $i \in N$. Since G is better-reply closed relative to u and $(x, u^*) \in \mathrm{cl}(\mathrm{graph}(u))$, this implies x is a Nash equilibrium of G. ∎

5.2   Two Characterization of the Limit Problem for ε-Equilibria

The proof of Theorem 5.1 shows that better-reply closedness of a game relative to its payoff function (and, therefore, all the conditions that are stronger than that) easily yield a limit result. Thus, this condition is particularly useful to establish the existence of Nash equilibria via the existence of approximate equilibria. But, such condition is only useful whenever the remaining assumptions of Theorem 5.1 are satisfied. Of these, two important ones are the existence of a sequence $\{f_k\}_{k=1}^{\infty}$ such that $f_k \in F(G_k)$ for all $k \in \mathbb{N}$ and $\liminf_k f_i^k(x_k) \geq v_i(x)$ for all $i \in N$ and the existence of a sequence $\{x_k\}_{k=1}^{\infty} \subseteq X$ such that x_k is an f_k-equilibrium of G_k for all $k \in \mathbb{N}$.

An easy way of obtaining the former condition is to focus on ε-equilibria. In fact, in this case, we have that $f_k = v - 1/k$ for all $k \in \mathbb{N}$. Focusing on ε-equilibria is also appealing because, in this case, the limit problem (i.e. the problem of finding when a statement like Theorem 5.1 is valid) is easily characterized.

Such characterization will be obtained using the following notion. We say that a normal-form game $G = (X_i, u_i)_{i \in N}$ is *sequential better-reply secure* if whenever $\{(x_k, u(x_k))\}_{k=1}^{\infty} \subseteq X \times \mathbb{R}^n$ is a convergent sequence with limit $(x, \gamma) \in X \times \mathbb{R}^n$ and x is not a Nash equilibrium of G, there exists $i \in N$, $\eta > \gamma_i$, a subsequence $\{x_l\}_{l=1}^{\infty}$ of $\{x_k\}_{k=1}^{\infty}$ and a sequence $\{y_i^l\}_{l=1}^{\infty} \subseteq X_i$ such that $u_i(y_i^l, x_{-i}^l) \geq \eta$ for all $l \in \mathbb{N}$.

We consider two limit problems. The first one is more restrictive in the sense that we do not allow for a game to be approximated. We say that a normal-form game $G = (X_i, u_i)_{i \in N}$ has the *weak limit property* if, whenever $x \in X$, $\{x_k\}_{k=1}^{\infty} \subseteq X$ and $\{\varepsilon_k\}_{k=1}^{\infty} \subseteq \mathbb{R}_{++}$ are such that x_k is an ε_k-equilibrium of G for all $k \in \mathbb{N}$, $\lim_k x_k = x$ and $\lim_k \varepsilon_k = 0$, then x is a Nash equilibrium of G.

The second limit problem drops the above restrictions. We say that a normal-form game $G = (X_i, u_i)_{i \in N}$ has the *strong limit property* if, whenever $x \in X$, $\{x_k\}_{k=1}^{\infty} \subseteq X$, $\{\varepsilon_k\}_{k=1}^{\infty} \subseteq \mathbb{R}_{++}$ and $\{G_k\}_{k=1}^{\infty} \subseteq \mathbb{G}_s(X)$ are such that x_k is an ε_k-equilibrium of G_k for all $k \in \mathbb{N}$, $\lim_k x_k = x$, $\lim_k \varepsilon_k = 0$ and $\lim_k d(G_k, G) = 0$, then x is a Nash equilibrium of G.

Theorem 5.2 shows that the above conditions are equivalent.

Theorem 5.2   *Let $G = (X_i, u_i)_{i \in N} \in \mathbb{G}$. Then the following conditions are equivalent:*

1. *G is sequentially better-reply secure.*
2. *G has the strong limit property.*
3. *G has the weak limit property.*

Proof.   We first show that condition 1 implies condition 2. Let $x \in X$, $\{x_k\}_{k=1}^{\infty} \subseteq X$, $\{\varepsilon_k\}_{k=1}^{\infty} \subseteq \mathbb{R}_{++}$ and $\{G_k\}_{k=1}^{\infty}$ be such that x_k is an ε_k-equilibrium of G_k for all $k \in \mathbb{N}$, $\lim_k x_k = x$, $\lim_k \varepsilon_k = 0$ and $\lim_k d(G_k, G) = 0$. Suppose, in order to reach a contradiction, that x is not a Nash equilibrium of G.

Since u is bounded, we may assume that $\{u(x_k)\}_{k=1}^{\infty}$ converges. Let $\gamma = \lim_k u(x_k)$. Since G is sequentially better-reply secure, there exists $i \in N$, $\eta > \gamma_i$, a subsequence $\{x^l\}_{l=1}^{\infty}$ of $\{x^k\}_{k=1}^{\infty}$ and a sequence $\{y_i^l\}_{l=1}^{\infty} \subseteq X_i$ such that $u_i(y_i^l, x_{-i}^l) \geq \eta$ for all $l \in \mathbb{N}$.

We have that $u_i^l(x_l) \geq v_i^l(x_{-i}^l) - \varepsilon_l$ for all $i \in N$ and $l \in \mathbb{N}$ since x_l is an ε_l-equilibrium of G_l. Thus, for all $i \in N$ and $l \in \mathbb{N}$, $u_i(x_l) \geq v_i(x_{-i}^l) - \varepsilon_l - 2\|u_l - u\|$ and, therefore, $u_i(x_l) \geq \eta - \varepsilon_l - 2\|u_l - u\|$. Since $\lim_l \varepsilon_l = 0 = \lim_l \|u_l - u\|$, this implies that $\gamma_i = \lim_l u_i(x_l) \geq \eta$ for all $i \in N$, a contradiction. Hence, x is a Nash equilibrium of G and, thus, G has the strong limit property.

It is clear that every game that has the strong limit property has the weak limit property. Thus, condition 2 implies condition 3.

Finally, we show that condition 3 implies condition 1. Let $(x, \gamma) \in X \times \mathbb{R}^n$ and $\{(x_k, u(x_k))\}_{k=1}^{\infty} \subseteq X \times \mathbb{R}^n$ be such that $\lim_k (x_k, u(x_k)) = (x, \gamma)$ and x is not a Nash equilibrium. Define, for all $k \in \mathbb{N}$, $\varepsilon_k = \max_i(v_i(x_k) - u_i(x_k)) \geq 0$. Since payoff functions are bounded, we may assume $\{\varepsilon_k\}_{k=1}^{\infty}$ converges; let ε be its limit. Since x is not a Nash equilibrium of G and, by definition, x_k is an ε_k-equilibrium of G for all k, then the weak limit property of G implies that $\varepsilon > 0$.

Hence, the following holds for some k': For all $k \geq k'$, $\varepsilon_k > \varepsilon/2$, $u_j(x_k) > \gamma_j - \varepsilon/4$ for all $j \in N$ and there is $i \in N$ such that $v_i(x_k) > u_i(x_k) + \varepsilon/2$. Since there are finitely many players, there exist $i \in N$, a subsequence $\{x_l\}_{l=1}^{\infty}$ of $\{x_k\}_{k=1}^{\infty}$ and a sequence $\{y_i^l\}_{l=1}^{\infty} \subseteq X_i$ such that, for all $l \in \mathbb{N}$,

$$u_i(y_i^l, x_{-i}^l) > u_i(x_l) + \varepsilon/2 > \gamma_i + \varepsilon/4.$$

Hence, simply define $\eta = \gamma_i + \varepsilon/4$. This shows that G is sequentially better-reply secure. ∎

5.3 Sufficient Conditions for Limit Results

The characterization of the ε-equilibrium limit problem in terms of sequential better-reply security provided in Theorem 5.2 is useful, in particular, because it clarifies the logic of known limit results. Therefore, it facilitates the comparison between the different conditions introduced to address limit problems and, also, those introduced to address the related problem of existence of equilibrium.

We first consider generalized better-reply security and related conditions considered in Chapter 3 through the notion of very weak better-reply security. Let $G = (X_i, u_i)_{i \in N}$ be a normal-form game and $\tilde{u}$ be a bounded $\mathbb{R}^n$-valued function on X; furthermore, let $\tilde{v}_i(x_{-i}) = \sup_{x_i \in X_i} \tilde{u}_i(x_i, x_{-i})$

for all $i \in N$ and $x_{-i} \in X_{-i}$. We say that G is *very weakly better-reply secure relative to $\tilde{u}$* if

(a) $\tilde{u}_i \leq u_i$ for all $i \in N$,
(b) $\tilde{u}_i$ is weakly payoff secure for all $i \in N$, and
(c) x^* is a Nash equilibrium of G for all $(x^*, u^*) \in \mathrm{cl}(\mathrm{graph}(u))$ such that $u_i^* \geq \tilde{v}_i(x_{-i}^*)$.

Moreover, we say that G is *very weakly better-reply secure* if there exists a bounded $\mathbb{R}^n$-valued function $\tilde{u}$ on X such that G is very weakly better-reply secure relative to $\tilde{u}$.

Thus, very weak better-reply security is weaker than weak better-reply security because the former requires the better-behaved approximation $\tilde{u}$ only to be weakly payoff secure but not necessarily generalized payoff secure.

The two conditions differ also because the above definition does not require $\tilde{u}_i(\cdot, x_{-i})$ to be quasiconcave for all $i \in N$ and $x_{-i} \in X_{-i}$. This condition is not important for limit result and this is the reason why it is omitted; however, this condition would have added to the definition of very weak better-reply security if the focus would have been on the existence of equilibria.

The following result shows that very weak better-reply security is sufficient for sequential better-reply security.

Theorem 5.3   *If $G = (X_i, u_i)_{i \in N} \in \mathbb{G}$ is very weakly better-reply secure, then G is sequential better-reply secure.*

Proof.   Let $\tilde{u}$ be such that G is very weakly better-reply secure relative to $\tilde{u}$. Let $\{(x^k, u(x^k))\}_{k=1}^{\infty} \subseteq X \times \mathbb{R}^n$ be a convergent sequence with limit $(x, \gamma) \in X \times \mathbb{R}^n$ and such that x is not a Nash equilibrium of G.

Since G is very weakly better-reply secure relative to $\tilde{u}$ and $(x, \gamma) \in \mathrm{cl}(\mathrm{graph}(u))$, then there exists $i \in N$ such that $\tilde{v}_i(x_{-i}) > \gamma_i$. Since u_i is weakly payoff secure, then $\tilde{v}_i$ is lower semicontinuous. Letting $0 < \varepsilon < \tilde{v}_i(x_{-i}) - \gamma_i$ and $\eta = \tilde{v}_i(x_{-i}) - \varepsilon$, it follows that $\eta > \gamma_i$ and there exists a neighborhood V of x_{-i} such that $\tilde{v}_i(x'_{-i}) > \eta$ for all $x'_{-i} \in V$.

Let $K \in \mathbb{N}$ be such that $x_{-i}^k \in V$ for all $k \geq K$. Then, for all $k \geq K$, $\tilde{v}_i(x_{-i}^k) > \eta$ and, thus, there exists $y_i^k \in X_i$ such that $u_i(y_i^k, x_{-i}^k) \geq \tilde{u}_i(y_i^k, x_{-i}^k) > \eta$. This shows that G is sequentially better-reply secure. ∎

We obtain from Theorem 5.3 several sufficient conditions for a game to have the strong limit property.

Theorem 5.4 *Let $G \in \mathbb{G}$. Then G has the strong limit property if one of the following conditions hold:*

1. *G is generalized better-reply secure.*
2. *G is better-reply secure.*
3. *G is better-reply closed relative to u and weakly payoff secure.*

Proof. Suppose that G is generalized better-reply secure. Then, by Theorem 3.7, G is weakly better-reply secure and, hence, very weakly better-reply secure. Theorem 5.3 implies that G is sequentially better-reply secure and the conclusion follows from Theorem 5.2.

If G is better-reply secure, then G is generalized better-reply secure and, by the above, G has the strong limit property.

Finally, suppose that G is better-reply closed relative to u and weakly payoff secure. Hence, G is very weakly better-reply secure relative to u. Thus, G has the strong limit property by Theorems 5.2 and 5.3. ∎

The most general limit problem we have considered so far is the one in the definition of the strong limit property. The generality of this problem, as compared with the one in the weak limit property, arises because we allow for a sequence of games to approach a given (limit) game. Nevertheless, the notion of convergence of games is relatively strong since it requires players' payoff functions to converge uniformly and the players' strategy spaces to be equal to those in the limit game.

We next consider a weaker form of convergence of games, variational convergence. We first present a stronger version of this notion that facilitates the comparison with sequential better-reply security and, therefore, with the weak and strong limit properties.

Let $G = (X_i, u_i)_{i \in N} \in \mathbb{G}$. A sequence of games $G_k = (X_i^k, u_i^k)_{i \in N} \in \mathbb{G}(X)$ *strongly variationally converges* to G if:

(A) The following property holds for all $i \in N$, $\varepsilon > 0$, $x_i \in X_i$ and $\tilde{x} \in \mathrm{Ls}(X_k)$: If $\{G_j\}_{j=1}^{\infty}$ is a subsequence of $\{G_k\}_{k=1}^{\infty}$ and $\{\tilde{x}_{-i}^j\}_{j=1}^{\infty}$ is such that $\lim_j \tilde{x}_{-i}^j = \tilde{x}_{-i}$ and $\tilde{x}_{-i}^j \in X_{-i}^j$ for all $j \in \mathbb{N}$, then there exists $J \in \mathbb{N}$ such that $v_i^j(\tilde{x}_{-i}^j) > u_i(x_i, \tilde{x}_{-i}) - \varepsilon$ for all $j \geq J$, and

(B) For all $(x, \alpha) \in \mathrm{Ls}(\mathrm{graph}(u_k)) \backslash \mathrm{graph}(u)$, there exists $i \in N$ such that $v_i(x_{-i}) > \alpha$.

An alternative but related notion of convergence of games is the following. Given a game $G = (X_i, u_i)_{i \in N} \in \mathbb{G}$, we say that a sequence of games $G_k = (X_i^k, u_i^k)_{i \in N} \in \mathbb{G}(X)$ is *multi-hypoconvergent* to G if for all $i \in N$

and $x \in X$ the following conditions hold for every subsequence $\{G_j\}_{j=1}^{\infty}$ of $\{G_k\}_{k=1}^{\infty}$:

(a) If $\{x_{-i}^j\}_{j=1}^{\infty}$ is such that $x_{-i}^j \in X_{-i}^j$ for all $j \in \mathbb{N}$ and $\lim_j x_{-i}^j = x_{-i}$, then there exists $\{x_i^j\}_{j=1}^{\infty}$ such that $x_i^j \in X_i^j$ for all $j \in \mathbb{N}$, $\lim_j x_i^j = x_i$ and $\liminf_j u_i^j(x_i^j, x_{-i}^j) \geq u_i(x)$;

(b) If $\{x_j\}_{j=1}^{\infty}$ is such that $x_j \in X_j$ for all $j \in \mathbb{N}$ and $\lim_j x_j = x$, then $\limsup_j u_i^j(x_j) \leq u_i(x)$.

A third way of approximating a game that we consider is as follows. We say that $G = (X_i, u_i)_{i \in N} \in \mathbb{G}$ is *strongly function approximated by* $\{G_k\}_{k=1}^{\infty}$, where $G_k = (X_i, u_i^k)_{i \in N} \in \mathbb{G}(X)$ for all $k \in \mathbb{N}$, if the following conditions hold:

(I) For all sequences $\{x_k\}_{k=1}^{\infty} \subseteq X$ and $\varepsilon > 0$, there exists $K \in \mathbb{N}$ such that $u_k(x_k) \leq u(x_k) + \varepsilon$ for all $k \geq K$.

(II) For all $x \in X$ and $\{x_k\}_{k=1}^{\infty} \subseteq X$ such that $\lim_k x_k = x$, $\liminf_k u_i^k(x_k) \geq v_i(x_{-i})$ for all $i \in N$.

In order to relate the above conditions to sequential better-reply security, the latter needs to be extended. In fact, the definition of sequential better-reply security presented before is one that makes no reference to approximating sequences of games; in contrast, both strong variational convergence and multi-hypoconvergence refer to a specific sequence of games. Given $G = (X_i, u_i)_{i \in N} \in \mathbb{G}$ and $\{G_k\}_{k=1}^{\infty} \in \mathbb{G}(X)$, where $G_k = (X_i^k, u_i^k)_{i \in N}$ for all $k \in \mathbb{N}$, we say that G is *sequentially better-reply secure with respect to* $\{G_k\}_{k=1}^{\infty}$ if for all subsequences $\{G_j\}_{j=1}^{\infty}$ of $\{G_k\}_{k=1}^{\infty}$ and all convergent sequences $\{(x_j, u_j(x_j))\}_{j=1}^{\infty}$ such that $x_j \in X_j$ for all $j \in \mathbb{N}$ and $\lim_j x_j$ is not a Nash equilibrium of G, there exist $i \in N$, $\eta > \lim_j u_i^j(x_j)$, a subsequence $\{x_l\}_{l=1}^{\infty}$ of $\{x_j\}_j$ and a sequence $\{y_i^l\}_{l=1}^{\infty}$ such that $y_i^l \in X_i^l$ and $u_i^l(y_i^l, x_{-i}^l) \geq \eta$ for all $l \in \mathbb{N}$.

Similarly to Theorem 5.2, we obtain that the sequential better-reply security of a game G with respect to a sequence of games $\{G_k\}_{k=1}^{\infty}$ is both sufficient and necessary for accumulation points of ε-equilibria of the games in this sequence to be Nash equilibria of G.

Theorem 5.5 *Let $G = (X_i, u_i)_{i \in N} \in \mathbb{G}$ and $\{G_k\}_{k=1}^{\infty} \subseteq \mathbb{G}(X)$ where $G_k = (X_i^k, u_i^k)_{i \in N}$ for all $k \in \mathbb{N}$. Then, the following conditions are equivalent:*

1. *G is sequentially better-reply secure with respect to $\{G_k\}_{k=1}^{\infty}$.*

2. *If $\{G_{k_j}\}_{j=1}^{\infty}$ is a subsequence of $\{G_k\}_{k=1}^{\infty}$, $x \in X$ and $\{(x_j,\varepsilon_j)\}_{j=1}^{\infty}$ is such that $(x_j,\varepsilon_j) \in X_j \times \mathbb{R}_+$ and x_j is an ε_j-equilibrium of G_j for all $j \in \mathbb{N}$, $\lim_j(x_j,\varepsilon_j) = (x,0)$ and $\{u_j(x_j)\}_{j=1}^{\infty}$ converges, then x is a Nash equilibrium of G.*

Proof. We first show that condition 1 implies condition 2. Let $\{G_j\}_{j=1}^{\infty}$ be a subsequence of $\{G_k\}_{k=1}^{\infty}$, $x \in X$ and $\{(x_j,\varepsilon_j)\}_{j=1}^{\infty}$ be such that $(x_j,\varepsilon_j) \in X_j \times \mathbb{R}_+$ and x_j is an ε_j-equilibrium of G_j for all $j \in \mathbb{N}$, $\lim_j(x_j,\varepsilon_j) = (x,0)$ and $\{u_j(x_j)\}_{j=1}^{\infty}$ converges. Suppose, in order to reach a contradiction, that x is not a Nash equilibrium of G.

Let $\gamma = \lim_j u_j(x_j)$. Since G is sequentially better-reply secure with respect to $\{G_k\}_{k=1}^{\infty}$, there exists $i \in N$, $\eta > \gamma_i$, a subsequence $\{x_l\}_{l=1}^{\infty}$ of $\{x_j\}_{j=1}^{\infty}$ and a sequence $\{y_i^l\}_{l=1}^{\infty}$ such that, for all $l \in \mathbb{N}$, $y_i^l \in X_i^l$ and $u_i^l(y_i^l, x_{-i}^l) \geq \eta$.

Since x_l is a ε_l-equilibrium of G_l, then $u_i^l(x_l) \geq v_i^l(x_{-i}^l) - \varepsilon_l \geq \eta - \varepsilon_l$ for all $i \in N$ and $l \in \mathbb{N}$. Since $\lim_l \varepsilon_l = 0$, this implies that $\gamma_i = \lim_l u_i^l(x_l) \geq \eta$ for all $i \in N$, a contradiction. Hence, x is a Nash equilibrium of G and, thus, condition 2 holds.

Finally, we show that condition 2 implies condition 1. Let $\{G_j\}_{j=1}^{\infty}$ be a subsequence of $\{G_k\}_{k=1}^{\infty}$, $\{x_j, u^j(x_j)\}_{j=1}^{\infty} \subseteq X \times \mathbb{R}^n$ converge to (x,γ) and $x \notin E(G)$.

Define, for all $j \in \mathbb{N}$, $\varepsilon_j = \max_i(v_i^j(x_j) - u_i^j(x_j)) \geq 0$. Since $x \notin E(G)$ and x_j is an ε_j-equilibrium of G_j for all $j \in \mathbb{N}$, then condition 2 implies that $\{\varepsilon_j\}_{j=1}^{\infty}$ does not converge to 0.

Hence, there exist $\varepsilon > 0$ and a subsequence $\{x_l\}_{l=1}^{\infty}$ such that, for all $l \in \mathbb{N}$, $\varepsilon_l \geq \varepsilon$ and $u_i^l(x_l) > \gamma_i - \varepsilon/4$ for all $i \in N$. Thus, for all $l \in \mathbb{N}$, there is $i \in N$ such that $v_i^l(x_l) \geq u_i^l(x_l) + \varepsilon$. Since N is finite, there exist $i \in N$, a subsequence $\{x^m\}_{m=1}^{\infty}$ of $\{x_l\}_{l=1}^{\infty}$ and a sequence $\{y_i^m\}_{m=1}^{\infty}$ such that, for all $m \in \mathbb{N}$, with $y_i^m \in X_i^m$ and

$$u_i^m(y_i^m, x_{-i}^m) > u_i^m(x_m) + \varepsilon/2 > \gamma_i + \varepsilon/4.$$

Hence, simply define $\eta = \gamma_i + \varepsilon/4$. This shows that condition 1 holds. ∎

We next provide sufficient conditions for the sequential better-reply security of a game G relative to a sequence $\{G_k\}_{k=1}^{\infty}$ in terms of the convergence notions introduced previously.

Theorem 5.6 *Let $G = (X_i, u_i)_{i \in N} \in \mathbb{G}$ and, for all $k \in \mathbb{N}$, $G_k = (X_i^k, u_i^k)_{i \in N} \in \mathbb{G}(X)$. Then the following holds:*

1. *If $\{G_k\}_{k=1}^{\infty}$ strongly variationally converges to G then G is sequentially better-reply secure with respect to $\{G_k\}_{k=1}^{\infty}$.*
2. *If G is weakly reciprocal upper semicontinuous and strongly function approximated by $\{G_k\}_{k=1}^{\infty}$ then $\{G_k\}_{k=1}^{\infty}$ strongly variationally converges to G.*
3. *If $\{G_k\}_{k=1}^{\infty}$ is multi-hypoconvergent to G then $\{G_k\}_{k=1}^{\infty}$ strongly variationally converges to G.*

Proof.   We start with the proof of part 1. Let $\{G_j\}_{j=1}^{\infty}$ be a subsequence of $\{G_k\}_{k=1}^{\infty}$ and $\{x_j, u_j(x_j)\}_{j=1}^{\infty}$ be a convergent sequence such that $x_j \in X_j$ for all $j \in \mathbb{N}$ and $\lim_j x_j$ is not a Nash equilibrium of G. Let $(x, \gamma) = \lim_j (x_j, u_j(x_j))$ and note that $(x, \gamma) \in \mathrm{Ls}(\mathrm{graph}(u_k))$.

If $(x, \gamma) \notin \mathrm{graph}(u)$, then it follows by condition (B) in the definition of the strong variational convergence that there is $i \in N$ and $\bar{x}_i \in X_i$ such that $u_i(\bar{x}_i, x_{-i}) > \gamma_i$. The same conclusion holds when $(x, \gamma) \in \mathrm{graph}(u)$. Indeed, since $x \notin E(G)$, then there exists $i \in N$ and $\bar{x}_i \in X_i$ such that $u_i(\bar{x}_i, x_{-i}) > u_i(x)$. Since $(x, \gamma) \in \mathrm{graph}(u)$, then $u_i(x) = \gamma_i$ and, therefore, $u_i(\bar{x}_i, x_{-i}) > \gamma_i$.

Hence, let $i \in N$ and $\bar{x}_i \in X_i$ be such that $u_i(\bar{x}_i, x_{-i}) > \gamma_i$. Furthermore, let $\varepsilon > 0$ be such that $u_i(\bar{x}_i, x_{-i}) > \gamma_i + 2\varepsilon$. It then follows by condition (A) in the definition of the strong variational convergence that there exists $\{x_i^j\}_{j=1}^{\infty}$ and $J \in \mathbb{N}$ such that, for all $j \geq J$, $u_i^j(x^j) > u_i(\bar{x}_i, x_{-i}) - \varepsilon > \gamma_i + \varepsilon$. Hence, simply let $\eta = \gamma_i + \varepsilon$. This shows that G is sequentially better-reply secure with respect to $\{G_k\}_{k=1}^{\infty}$.

We turn to the proof of part 2. Let G be weakly reciprocal upper semicontinuous and strongly function approximated by $\{G_k\}_{k=1}^{\infty}$. We claim that $\{G_k\}_{k=1}^{\infty}$ strongly variationally converges to G.

Regarding part (A) of the definition of the strong variational convergence, let $i \in N$, $\varepsilon > 0$, $x_i \in X_i$, $\tilde{x} \in \mathrm{Ls}(X_k)$, $\{G_{k_j}\}_{j=1}^{\infty}$ be a subsequence of $\{G_k\}_{k=1}^{\infty}$ and $\{\tilde{x}_{-i}^{k_j}\}_{j=1}^{\infty}$ be such that $\lim_j \tilde{x}_{-i}^{k_j} = \tilde{x}_{-i}$ and $\tilde{x}_{-i}^{k_j} \in X_{-i}^{k_j}$ for all $j \in \mathbb{N}$.

Let $\{\hat{x}_l\}_{l=1}^{\infty}$ be such that $\lim_l \hat{x}_l = \tilde{x}$ and define $\{x_l\}_{l=1}^{\infty}$ by

$$
x_l = \begin{cases} (x_i, \tilde{x}_{-i}^{k_j}) & \text{if there exists } j \in \mathbb{N} \text{ such that } k_j = l, \\ \hat{x}_l & \text{otherwise.} \end{cases}
$$

We have that $\liminf_j u_{k_j}(x_i, x_{-i}^{k_j}) \geq \liminf_l u_l(x_l)$. Hence, by condition (II) in the definition of strong function approximation, it follows that $\liminf_j u_{k_j}(x_i, x_{-i}^{k_j}) \geq v_i(\tilde{x}_{-i})$ for all $i \in N$. Thus, for all $\varepsilon > 0$, there exists

$J \in \mathbb{N}$ such that $v_i^{k_j}(x_{-i}^{k_j}) \geq u_i^{k_j}(x_i, x_{-i}^{k_j}) > v_i(\tilde{x}_{-i}) - \varepsilon \geq u_i(x_i, \tilde{x}_{-i}) - \varepsilon$ for all $j \geq J$.

Regarding part (B) of the definition of the strong variational convergence, let $(x, \alpha) \in \mathrm{Ls}(\mathrm{graph}(u_k)) \setminus \mathrm{graph}(u)$. Since u is bounded, we may assume that $\{u(x_k)\}_{k=1}^{\infty}$ converges. Let $\gamma = \lim_k u(x_k)$. Then condition (I) of the definition of strong function approximation implies that $\gamma \geq \alpha$.

We consider two cases. The first is when $(x, \gamma) \in \mathrm{graph}(u)$. In this case, it follows that $\alpha \neq \gamma$ since otherwise we would have $(x, \alpha) \in \mathrm{graph}(u)$. Since $\gamma \geq \alpha$, there exists $i \in N$ such that $\gamma_i > \alpha_i$ and, therefore, $v_i(x_{-i}) \geq u_i(x) = \gamma_i > \alpha_i$.

The second case is when $(x, \gamma) \notin \mathrm{graph}(u)$. In this case, since G is weakly reciprocal upper semicontinuous, there exists $i \in N$ such that $v_i(x_{-i}) > \gamma_i \geq \alpha_i$.

Thus, condition (B) in the definition of the strong variational convergence holds in both cases. Since condition (A) in the definition of the strong variational convergence also holds, then it follows that $\{G_k\}_{k=1}^{\infty}$ strongly variationally converges to G.

We finally establish part 3. Suppose that $\{G_k\}_{k=1}^{\infty}$ is multi-hypoconvergent to G. We claim that $\{G_k\}_{k=1}^{\infty}$ strongly variationally converges to G.

Regarding part (A) of the definition of the strong variational convergence, let $i \in N$, $\varepsilon > 0$, $x_i \in X_i$, $\tilde{x} \in \mathrm{Ls}(X_k)$, $\{G_j\}_{j=1}^{\infty}$ be a subsequence of $\{G_k\}_{k=1}^{\infty}$ and $\{\tilde{x}_{-i}^j\}_{j=1}^{\infty}$ be such that $\lim_j \tilde{x}_{-i}^j = \tilde{x}_{-i}$ and $\tilde{x}_{-i}^j \in X_{-i}^j$ for all $j \in \mathbb{N}$.

It follows by condition (a) in the definition of multi-hypoconvergence that there exists $\{\tilde{x}_i^j\}_{j=1}^{\infty}$ such that $\liminf_j u_i^j(\tilde{x}_j) \geq u_i(\tilde{x})$. Hence, for all $\varepsilon > 0$, there exists $J \in \mathbb{N}$ such that $\inf_{j \geq J} u_i^j(\tilde{x}_j) > u_i(\tilde{x}) - \varepsilon$. This establishes part (A) of the definition of the strong variational convergence.

Regarding part (B) of the definition of the strong variational convergence, let $(x, \alpha) \in \mathrm{Ls}(\mathrm{graph}(u_k)) \setminus \mathrm{graph}(u)$. Let $\{x_j\}_{j=1}^{\infty}$ be such that $x_j \in X_j$ for all $j \in \mathbb{N}$ and $\lim_j (x_j, u^j(x_j)) = (x, \gamma)$. It follows by part (b) of the definition of multi-hypo convergence that $\gamma_i \leq u_i(x)$ for all $i \in N$. Since $(x, \gamma) \notin \mathrm{graph}(u)$, then there exists $i \in N$ such that $u_i(x) > \gamma_i$. This proves part (B) of the definition of the strong variational convergence.

Since parts (A) and (B) of the definition of the strong variational convergence hold, it follows that $\{G_k\}_{k=1}^{\infty}$ strongly variationally converges to G. ∎

The conditions considered above can be weakened by requiring them to holds only along sequences of Nash equilibria. We illustrate this by defining a weaker version of sequential better-reply security of a game with respect to a sequence of games.

Given $G = (X_i, u_i)_{i \in N} \in \mathbb{G}$ and $\{G_k\}_{k=1}^{\infty} \in \mathbb{G}(X)$, where $G_k = (X_i^k, u_i^k)_{i \in N}$ for all $k \in \mathbb{N}$, we say that G is *very weakly sequential better-reply secure with respect to* $\{G_k\}_{k=1}^{\infty}$ if for all subsequences $\{G_j\}_{j=1}^{\infty}$ of $\{G_k\}_{k=1}^{\infty}$ and all convergent sequences $\{(x_j, u_j(x_j))\}_{j=1}^{\infty}$ such that $x_j \in E(G_j)$ for all $j \in \mathbb{N}$ and $\lim_j x_j$ is not a Nash equilibrium of G, there exist $i \in N$, $\eta > \lim_j u_i^j(x_j)$, a subsequence $\{x_l\}_{l=1}^{\infty}$ of $\{x_j\}_{j=1}^{\infty}$ and a sequence $\{y_i^l\}_{l=1}^{\infty}$ such that $y_i^l \in X_i^l$ and $u_i^l(y_i^l, x_{-i}^l) \geq \eta$ for all $l \in \mathbb{N}$.

Using a similar argument to that of Theorem 5.5, we can show that the weak sequential better-reply security of a game G with respect to $\{G_k\}_{k=1}^{\infty}$ is equivalent to the following limit condition: If $\{G_j\}_{j=1}^{\infty}$ is a subsequence of $\{G_k\}_{k=1}^{\infty}$, $x \in X$ and $\{x_j\}_{j=1}^{\infty}$ is such that $x_j \in E(G_j)$ for all $j \in \mathbb{N}$, $\lim_j x_j = x$ and $\{u_j(x_j)\}_{j=1}^{\infty}$ converges, then x is a Nash equilibrium of G. Thus, each limit point of Nash equilibria of a sequence of games approximating G in the sense of weak sequential better-reply security is a Nash equilibria. However, note that weak sequential better-reply security is too weak to imply the conclusion of Theorem 5.5, i.e., for sequences of ε-equilibria with ε converging to zero.

Furthermore, a result analogous to Theorem 5.6 holds when strong variational convergence, strong function approximation and weak reciprocal upper semicontinuity have been replaced with analogous weaker versions.

A final variation we can consider is to weaken sequential better-reply security as above. We say that a normal-form game $G = (X_i, u_i)_{i \in N}$ is *weakly sequential better-reply secure* if whenever $\{(x_k, u(x_k))\}_{k=1}^{\infty} \subseteq X \times \mathbb{R}^n$ is a convergent sequence with limit $(x, \gamma) \in X \times \mathbb{R}^n$, $x_k \in E(G)$ for all $k \in \mathbb{N}$ and x is not a Nash equilibrium of G, there exists $i \in N$, $\eta > \gamma_i$, a subsequence $\{x^l\}_{l=1}^{\infty}$ of $\{x^k\}_{k=1}^{\infty}$ and a sequence $\{y_i^l\}_{l=1}^{\infty} \subseteq X_i$ such that $u_i(y_i^l, x_{-i}^l) \geq \eta$ for all $l \in \mathbb{N}$.

A result analogous to Theorem 5.2 holds for weak sequential better-reply security when both the weak and strong limit properties are weakened by having $\varepsilon_k = 0$ for all $k \in \mathbb{N}$. In the case of the weak limit property, this change simply means that the set of Nash equilibria is closed.

As a consequence of the above discussion, it follows that multi-player well-behaved security and generalized C-security are examples of sufficient conditions for weak sequential better-reply security. However, in general, these conditions are not sufficient for sequential better-reply security as Example 4.13 illustrates.

Recall that the game $G = (X_i, u_i)_{i \in N} \in \mathbb{G}$ in Example 4.13 is such that $N = \{1, 2\}$, $X_1 = X_2 = [0, 1]$,

$$u_1(x_1, x_2) = \begin{cases} 1 & \text{if } x_1 = 1, \\ 0 & \text{if } x_1 < 1 \text{ and } x_2 = 1/2, \\ 2x_2 & \text{if } x_1 < 1 \text{ and } x_2 < 1/2, \\ 2(1 - x_2) & \text{if } x_1 < 1 \text{ and } x_2 > 1/2, \end{cases}$$

and $u_2(x_1, x_2) = u_1(x_2, x_1)$ for all $(x_1, x_2) \in X$.

We have shown that G is generalized C-secure. We now show that G does not have the weak limit property. Indeed, consider $x_k = (1/2 - 1/(2k), 1/2 - 1/(2k))$ for all $k \in \mathbb{N}$. Then, $\lim_k x_k = (1/2, 1/2)$ and x_k is a $1/k$-equilibrium for all $k \in \mathbb{N}$. Indeed, for all $i \in N$ and $k \in \mathbb{N}$, $u_i(x_k) = 1 - 1/k$ and $v_i \equiv 1$. But $x^* = (1/2, 1/2)$ is not a Nash equilibrium of G.

5.4 Existence of ε-Equilibrium

We consider the existence of ε-equilibria in this section. While the focus of this book is on the existence of Nash equilibria (i.e. of ε-equilibria with $\varepsilon = 0$), the existence of ε-equilibria for $\varepsilon > 0$ is interesting for at least two reasons. First, some games have ε-equilibria for all $\varepsilon > 0$ but not Nash equilibria (see Prokopovych (2011) for an auction example). And, second, as a tool to address the existence of Nash equilibria via the limit results of the previous section.

Existence of approximate equilibria was already considered in Lemma 3.5. This result considers the game $\underline{G} = (X_i, \underline{u}_i)_{i \in N}$ derived from a given game $G = (X_i, u_i)_{i \in N}$ by replacing the payoff function of each player by its generalized payoff secure envelope. Furthermore, it requires the target function f to be continuous. As the next result shows these properties, generalized payoff security and continuity of the target function, which in the case of ε-equilibria equals $v - \varepsilon$, are sufficient for the existence of ε-equilibrium for all $\varepsilon > 0$.

Theorem 5.7 *If $G = (X_i, u_i)_{i \in N} \in \mathbb{G}_q$ is generalized payoff secure and weakly continuous, then G has an ε-equilibrium for all $\varepsilon > 0$.*

Proof. Let $G = (X_i, u_i)_{i \in N} \in \mathbb{G}_q$ be generalized payoff secure and weakly continuous. Then v_i is continuous for all $i \in N$ and, by Theorem 3.17, $u = \underline{u}$. Clearly, it follows that $v = \underline{v}$.

Let $\varepsilon > 0$ and define $f : X \to \mathbb{R}^n$ by $f_i(x) = v_i(x_{-i}) - \varepsilon$ for all $i \in N$ and $x \in X$. Then, f is continuous and such that $f \in F(\underline{G})$, $f_i(x) = f_i(x'_i, x_{-i})$ and $f_i(x) < \underline{v}_i(x_{-i})$ for all $i \in N$, $x'_i \in X_i$ and $x \in X$.

It then follows by Lemma 3.5 that $\underline{G}$ has a f-equilibrium $x^* \in X$. Then, for all $i \in N$, $u_i(x^*) = \underline{u}_i(x^*) \geq f_i(x^*) = v_i(x^*_{-i}) - \varepsilon$ and we conclude that x^* is an ε-equilibrium of G. ∎

We next consider an alternative approach to establish the existence of ε-equilibria. Such approach is similar to the one considered in Theorems 3.26 and 3.27. In fact, as compared with the previous result, both the quasiconcavity and the upper semicontinuity requirements will be strengthened while the lower semicontinuity ones will be dropped.

Recall that a polytope is the convex hull of finitely many points in an Euclidian space. Let $G = (X_i, u_i)_{i \in N} \in \mathbb{G}_q$ be such that X_i is a polytope for all $i \in N$. For all $i \in N$, we say that player i's payoff function is *piecewise quasiconcave* if there is a finite cover of X_{-i} by compact convex subsets X^l_{-i} such that u_i is quasiconcave on $X_i \times X^l_{-i}$ for all l. Furthermore, we say that player i's payoff function is *i-upper semicontinuous* if

$$\limsup_k v_i(\alpha_k z^k_{-i} + (1 - \alpha_k)x_{-i}) \leq \limsup_k v_i(z^k_{-i}) \qquad (5.1)$$

for all $x_{-i} \in X_{-i}$, $\{z^k_{-i}\}_{k=1}^{\infty}$ converging to x_{-i} and $\{\alpha_k\}_{k=1}^{\infty} \subseteq (0, 1]$ converging to zero. Clearly, i-upper semicontinuity is implied by the continuity of v_i and so can be thought of as a weak form of continuity. We note, however, that i-upper semicontinuity is neither implied nor does it imply upper semicontinuity (see footnote 4 in Carmona (2010) for details). As Theorem 5.11 below will show, i-upper semicontinuity is better understood as a quasiconcavity requirement rather than an upper semicontinuity one.

The existence of ε-equilibrium follows from the above assumptions as the next result shows.

Theorem 5.8   *Let $G = (X_i, u_i)_{i \in N} \in \mathbb{G}_q$ be such that, for all $i \in N$, X_i is a polytope and u_i is upper semicontinuous, i-upper semicontinuous and piecewise quasiconcave. Then G has an ε-equilibrium for all $\varepsilon > 0$.*

The proof of Theorem 5.8 relies on two lemmas. The first shows that if P is a polytope then, for some $r > 0$, the ball of radius r intersected with the cone of all interior directions from x is a subset of P.

Lemma 5.9   *Let $P \subseteq \mathbb{R}^n$ be a polytope. Then, for all $x \in P$, there exists $r > 0$ such that for all $\tilde{x} \in P$, $\tilde{x} \neq x$,*

$$x + r \frac{\tilde{x} - x}{\|\tilde{x} - x\|} \in P.$$

Proof.   Note that polytopes are intersections of finitely many half-spaces. Let $P = \cap_{i \in F} \{z \in \mathbb{R}^n : c_i \cdot z \leq d_i\}$ for some finite set F, vectors $c_i \in \mathbb{R}^n$ and scalars $d_i \in \mathbb{R}$. Let $T = \{i \in F : c_i \cdot x = d_i\}$ and $T^c = F \backslash T$. Take r to be the smallest distance of x from the hyperplanes $H_i = \{z \in \mathbb{R}^n : c_i \cdot z = d_i\}$ for all $i \in T^c$, unless $T^c = \emptyset$, in which case we can take $r = 1$.

Let $\tilde{x} \in P$ and define $\bar{x} = x + r(\tilde{x} - x)/\|\tilde{x} - x\|$. We claim that $c_i \cdot \bar{x} \leq d_i$ holds for all $i \in F$, which clearly implies that $\bar{x} \in P$. For convenience, for all $i \in F$, let $S_i = \{z \in \mathbb{R}^n : c_i \cdot z \leq d_i\}$ and $d(x, H_i) = \inf_{z \in H_i} \|z - x\|$ denote the distance of x from H_i. In the case $i \in T$, we have that $c_i \cdot x = d_i$ and $c_i \cdot \tilde{x} \leq d_i$ and so $c_i \cdot \bar{x} = d_i + r(c_i \cdot \tilde{x} - d_i)/\|\tilde{x} - x\| \leq d_i + 0 = d_i$. In the case $i \in T^c$, note that $\|\bar{x} - x\| = r \leq d(x, H_i)$, which implies that $\bar{x} \in S_i$ for all $i \in F$. In fact, if $\bar{x} \notin S_i$ for some $i \in F$, then define $\theta = (d_i - c_i \cdot x)/(c_i \cdot \bar{x} - c_i \cdot x) \in (0, 1)$ (since $0 < d_i - c_i \cdot x < c_i \cdot \bar{x} - c_i \cdot x$) and note that $\theta \bar{x} + (1 - \theta)x \in H_i$. But then we obtain that $\|\theta \bar{x} + (1 - \theta)x - x\| = \theta r < r \leq d(x, H_i)$, a contradiction. ∎

The following lemma is a simple consequence of Lemma 5.9 and will be used in the proof of both Theorem 5.8 and Theorem 5.11.

Lemma 5.10   *Let P be a polytope, $x \in P$, $\{x_k\}_{k=1}^\infty$ and $\{\alpha_k\}_{k=1}^\infty$ be such that $\lim_k x_k = x$, $0 < \alpha_k \leq 1$ and $\alpha_k x_k + (1 - \alpha_k)x \in P$ for all $k \in \mathbb{N}$. Then, there exist sequences $\{\hat{x}_k\}_{k=1}^\infty$ and $\{\theta_k\}_{k=1}^\infty$ such that $\lim_k \theta_k = 0$, $\hat{x}_k \in P$, $\theta_k \in (0, 1)$ and $x_k = \theta_k \hat{x}_k + (1 - \theta_k)x$ for all k sufficiently large.*

Furthermore, if $\{\alpha_k\}_{k=1}^\infty$ is bounded away from zero, then $\{\hat{x}_k\}_{k=1}^\infty$ and $\{\theta_k\}_{k=1}^\infty$ can be chosen so that $\lim_k \hat{x}_k = x$.

Proof.   Note first that whenever $k \in \mathbb{N}$ is such that $x_k = x$ we can let $\hat{x}_k = x$ and $\theta_k = 1/k$. Thus, we may assume that $x_k \neq x$ for all $k \in \mathbb{N}$.

Let $r > 0$ be such that $x + r(\tilde{x} - x)/\|\tilde{x} - x\| \in P$ for all $\tilde{x} \in P$, $\tilde{x} \neq x$. For convenience, let $\bar{x}_k = \alpha_k x_k + (1 - \alpha_k)x$ for all k. Note that there exists $K \in \mathbb{N}$ such that $\|x_k - x\| < r$ for all $k \geq K$. For all k, define $\gamma_k = \|\bar{x}_k - x\|/r$ and $\hat{x}_k = (1/\gamma_k)\bar{x}_k + (1 - 1/\gamma_k)x = x + r(\bar{x}_k - x)/\|\bar{x}_k - x\|$. Clearly, $\hat{x}_k \in P$

for all k. Then, $\bar{x}_k = \gamma_k \hat{x}_k + (1 - \gamma_k)x$ and so $x_k = (\gamma_k/\alpha_k)\hat{x}_k + (1 - \gamma_k/\alpha_k)x$ for all k.

Thus, define $\theta_k = \gamma_k/\alpha_k$ for all k, which immediately implies that $x_k = \theta_k \hat{x}_k + (1 - \theta_k)x$ for all k. Furthermore, $\theta_k = \|\bar{x}_k - x\|/r\alpha_k = \|x_k - x\|/r \in (0,1)$ for all $k \geq K$ and $\lim_k \theta_k = 0$.

Finally, consider the case where $\{\alpha_k\}_{k=1}^{\infty}$ is bounded away from zero. In this case, let $K \in \mathbb{N}$ be such that $\|\bar{x}_k - x\|^{1/2} < \min\{r, 1\}$ for all $k \geq K$. For all k, define $\gamma_k = \|\bar{x}_k - x\|^{1/2}$ and, as before, $\hat{x}_k = (1/\gamma_k)\bar{x}_k + (1 - 1/\gamma_k)x$ and $\theta_k = \gamma_k/\alpha_k$. Clearly, $\theta_k > 0$ for all k and $\lim_k \theta_k = 0$, implying that $\theta_k \in (0,1)$ for sufficiently large k.

Note that $\|\hat{x}_k - x\| = \|\bar{x}_k - x\|/\gamma_k = \|\bar{x}_k - x\|^{1/2}$ and so $\lim_k \hat{x}_k = x$. It remains to show that $\hat{x}_k \in P$ for all $k \geq K$. Since $\|\bar{x}_k - x\|^{1/2} \leq r$, then $1/\|\bar{x}_k - x\|^{1/2} \leq r/\|\bar{x}_k - x\|$, which implies that $\hat{x}_k$ can be expressed as a convex combination of x and $x + r(\bar{x}_k - x)/\|\bar{x}_k - x\|$. Since the latter point belongs to P, then $\hat{x}_k$ also belongs to P. ∎

The following two special cases illustrate the usefulness of Lemma 5.10. The first case (to be considered in the proof of Theorem 5.8 below) is obtained by letting $\alpha_k = 1$ for all k. Thus, in this case, we have a sequence $\{x_k\}_{k=1}^{\infty}$ contained in the polytope P, converging to a point x also in P. Lemma 5.10 then asserts that, for all k sufficiently large, each x_k can be expressed as a convex combination between the limit point x and some other point $\hat{x}_k$ in P.

A second special case of interest (to be considered in the proof of Theorem 5.11 below) occurs when α_k converges to zero. In this case, we have a sequence $\{x_k\}_{k=1}^{\infty}$ converging to a point x in P with the property that $\alpha_k x_k + (1 - \alpha_k)x$ also belongs to P. Lemma 5.10 implies that x_k must itself belong to P for all k sufficiently large: In fact, for all k sufficiently large, each x_k can be expressed as a convex combination between the limit point x and some other point $\hat{x}_k$ in P.

We finally prove Theorem 5.8.

Proof of Theorem 5.8.   Let $\varepsilon > 0$, $i \in N$ and define $\beta_i : X_{-i} \rightrightarrows X_i$ by $\beta_i(x_{-i}) = \{x_i \in X_i : u_i(x_i, x_{-i}) > v_i(x_{-i}) - \varepsilon\}$ for all $x_{-i} \in X_{-i}$. We claim that β_i is lower hemicontinuous.

In order to establish that β_i is lower hemicontinuous, let $F \subseteq X_i$ be closed, $x_{-i} \in X_{-i}$, $\{x_{-i}^k\}_{k=1}^{\infty} \subseteq \{x'_{-i} \in X_{-i} : \beta_i(x'_{-i}) \subseteq F\}$ be such that $\lim_k x_{-i}^k = x_{-i}$ and $x_i \in \beta_i(x_{-i})$. We will show that $x_i \in F$, which implies that $\beta_i(x_{-i}) \subseteq F$. This, in turn, will show that β_i is lower hemicontinuous by Theorem A.7.

Let $0 < \eta < \varepsilon$ be such that $u_i(x_i, x_{-i}) > v_i(x_{-i}) - \varepsilon + \eta$. Since u_i is upper semicontinuous, so is v_i. Thus, let $\delta > 0$ be such that $v_i(x'_{-i}) < v_i(x_{-i}) + \eta/2$ for all $x'_{-i} \in X_{-i}$ satisfying $\|x_{-i} - x'_{-i}\| < \delta$. By Lemma 5.10 (with $\alpha_k = 1$ for all k), let $\{\hat{x}^k_{-i}\}_{k=1}^\infty$ and $\{\theta_k\}_{k=1}^\infty$ be such that $\lim_k \hat{x}^k_{-i} = x_{-i}$, $\lim_k \theta_k = 0$, $\hat{x}^k_{-i} \in X_i$, $\theta_k \in (0, 1)$ and $x^k_{-i} = \theta_k \hat{x}^k_{-i} + (1 - \theta_k) x_{-i}$ for all k sufficiently large. Taking a subsequence if necessary, we may assume that $\{v_i(\hat{x}^k_{-i})\}_{k=1}^\infty$ converges to $\limsup_k v_i(\hat{x}^k_{-i})$ and that $\hat{x}^k_{-i} \in X_i$, $\theta_k \in (0, 1)$ and $x^k_{-i} = \theta_k \hat{x}^k_{-i} + (1 - \theta_k) x_{-i}$ for all $k \in \mathbb{N}$.

Let $\hat{x}^k_i \in X_i$ be such that $u_i(\hat{x}^k_i, \hat{x}^k_{-i}) > v_i(\hat{x}^k_{-i}) - \varepsilon + \eta/2$ and define $x^k_i = \theta_k \hat{x}^k_i + (1 - \theta_k) x_i$ for all k. Since $\lim_k \theta_k = 0$, it follows that $\lim_k x^k_i = x_i$.

We claim that $x^k_i \in \beta_i(x^k_{-i})$ for all k sufficiently large. Suppose, in order to reach a contradiction, that there exists an infinite sequence $\{k_j\}_{j \in \mathbb{N}}$ such that $u_i(x^{k_j}_i, x^{k_j}_{-i}) \le v_i(x^{k_j}_{-i}) - \varepsilon$. Taking a further subsequence if needed, we may assume that there exists $l \in \{1, \ldots, L_i\}$ such that $\hat{x}^{k_j}_{-i} \in X^l_{-i}$ for all j. Since X^l_{-i} is compact, then $x_{-i} \in X^l_i$. Thus, the quasiconcavity of u_i in $X_i \times X^l_{-i}$ implies that

$$u_i(x^{k_j}_i, x^{k_j}_{-i}) \ge \min\{u_i(x_i, x_{-i}), u_i(\hat{x}^{k_j}_i, \hat{x}^{k_j}_{-i})\}$$
$$> \min\{v_i(x_{-i}) - \varepsilon + \eta, v_i(\hat{x}^{k_j}_{-i}) - \varepsilon + \eta/2\}$$
$$= v_i(\hat{x}^{k_j}_{-i}) - \varepsilon + \eta/2,$$

for all j such that $\|\hat{x}^{k_j}_{-i} - x_{-i}\| < \delta$ (since in this case, $v_i(\hat{x}^{k_j}_{-i}) < v_i(x_{-i}) + \eta/2$). This, together with $u_i(x^{k_j}_i, x^{k_j}_{-i}) \le v_i(x^{k_j}_{-i}) - \varepsilon$ for all $j \in \mathbb{N}$, implies that $v_i(\hat{x}^{k_j}_{-i}) + \eta/2 < v_i(x^{k_j}_{-i})$ for all j sufficiently large. Hence,

$$\limsup_k v_i(\hat{x}^k_{-i}) + \frac{\eta}{2} = \lim_j v_i(\hat{x}^{k_j}_{-i}) + \frac{\eta}{2}$$
$$\le \limsup_j v_i(x^{k_j}_{-i}) \le \limsup_k v_i(x^k_{-i}),$$

contradicting the i-upper semi-continuity of u_i. This contradiction establishes that $x^k_i \in \beta_i(x^k_{-i})$ for all k sufficiently large. Since $\beta_i(x^k_{-i}) \subseteq F$ for all k and F is closed, then $x_i = \lim_k x^k_i \in F$, as desired. Thus, β_i is lower hemicontinuous.

Let, for all $i \in N$, $B^\varepsilon_i : X_{-i} \rightrightarrows X_i$ be defined by $B^\varepsilon_i(x_{-i}) = \mathrm{cl}(\beta_i(x_{-i}))$ for all $x_{-i} \in X_{-i}$. Since β_i is lower hemicontinuous, it follows by Theorem A.7 that B^ε_i is lower hemicontinuous.

It is clear that $B_i^\varepsilon(x_{-i})$ is nonempty, closed and convex. Therefore, $B_\varepsilon(x) = B_1^\varepsilon(x_{-1}) \times \cdots \times B_n^\varepsilon(x_{-n})$ is also nonempty, closed and convex. Furthermore, $B_\varepsilon : X \rightrightarrows X$ is lower hemicontinuous. Thus, by Theorem A.16, there exists a continuous selection f of B_ε. Hence, by Theorem A.14, f has a fixed point x.

We claim that x is an ε-equilibrium of G. Let $i \in N$. Since x is a fixed point of f and f is a selection of B_ε, then $x_i \in B_i^\varepsilon(x_{-i}) = \text{cl}(\{x_i' \in X_i : u_i(x_i', x_{-i}) > v_i(x_{-i}) - \varepsilon\}$. Thus, there exists a sequence $\{x_i^k\}_{k=1}^\infty$ such that $\lim_k x_i^k = x_i$ and $u_i(x_i^k, x_{-i}) > v_i(x_{-i}) - \varepsilon$ for all $k \in \mathbb{N}$. Since u_i is upper semicontinuous, then $u_i(x_i, x_{-i}) > v_i(x_{-i}) - \varepsilon$. Since this holds for all $i \in N$, it follows that x is an ε-equilibrium of G. $\blacksquare$

We next seek sufficient conditions for i-upper semicontinuity. The following results shows that piecewise polyhedral concavity of the players' value functions implies i-upper semicontinuity.

Given $G = (X_i, u_i)_{i \in N}$ and $i \in N$, player i's value function is *piecewise polyhedral quasiconcave* if there is a finite cover of X_{-i} by compact convex subsets X_{-i}^l such that v_i is polyhedral quasiconcave on X_{-i}^l for all l.

Theorem 5.11   *Let $G = (X_i, u_i)_{i \in N} \in \mathbb{G}_q$ and $i \in N$. If v_i is piecewise polyhedral quasiconcave, then u_i is i-upper semicontinuous.*

Proof.   Let $x_{-i} \in X_{-i}$, $\{x_{-i}^k\}_{k=1}^\infty \subseteq X_{-i}$ be a sequence converging to x_{-i} and $\{\alpha_k\}_{k=1}^\infty \subseteq (0, 1]$ be a sequence converging to zero. Let $\bar{x}_{-i}^k = \alpha_k x_{-i}^k + (1 - \alpha_k)x_{-i}$ for all $k \in \mathbb{N}$.

Let $\gamma = \limsup_k v_i(\bar{x}_{-i}^k)$. Then, there exists a subsequence $\{\bar{x}_{-i}^{k_j}\}_{j=1}^\infty$ of $\{\bar{x}_{-i}^k\}_{k=1}^\infty$ such that $\lim_j v_i(\bar{x}_{-i}^{k_j}) = \gamma$. We may assume that $x_{-i}^{k_j} \neq x_{-i}$ for infinitely many indexes j, since otherwise $x_{-i}^{k_j} = \bar{x}_{-i}^{k_j} = x_{-i}$ for all j sufficiently large and so $\limsup_k v_i(x_{-i}^k) \geq \limsup_j v_i(x_{-i}^{k_j}) = v_i(x_{-i}) = \limsup_k v_i(\bar{x}_{-i}^k)$. Hence, taking a subsequence if necessary, we may assume that $x_{-i}^{k_j} \neq x_{-i}$ for all $j \in \mathbb{N}$; clearly, this implies that $\bar{x}_{-i}^{k_j} \neq x_{-i}$ for all $j \in \mathbb{N}$. Let $\varepsilon > 0$ and let $J_1 \in \mathbb{N}$ be such that $v_i(\bar{x}_{-i}^{k_j}) > \gamma - \varepsilon$ for all $j \geq J_1$.

Let $\{X_{-i}^l\}_{l=1}^{L_i}$ be a compact, convex cover of X_{-i} such that v_i is polyhedral quasiconcave on X_{-i}^l for all $l = 1, \ldots, L_i$. Since the cover is finite, we may assume that there exists $l \in \{1, \ldots, L_i\}$ such that $\bar{x}_{-i}^{k_j} \in X_{-i}^l$ for all $j \in \mathbb{N}$. Letting $P = \{y \in X_{-i}^l : v_i(y) \geq \gamma - \varepsilon\}$, then P is a polytope and $\bar{x}_{-i}^{k_j} \in P$ for all $j \geq J_1$. Furthermore $x_{-i} \in P$ since P is compact.

By Lemma 5.10, there exist $J > J_1$ and sequences $\{\hat{x}_{-i}^{k_j}\}_{k=1}^\infty$ and $\{\theta_{k_j}\}_{k=1}^\infty$ such that $\hat{x}_{-i}^{k_j} \in P$, $\theta_{k_j} \in (0, 1)$ and $x_{-i}^{k_j} = \theta_{k_j}\hat{x}_{-i}^{k_j} + (1 - \theta_{k_j})x_{-i}$ for all $j \geq J$. Hence, for all $j \geq J$, it follows that $x_{-i}^{k_j} \in P$, which implies that $v_i(x_{-i}^{k_j}) \geq \gamma - \varepsilon$ and so $\limsup_k v_i(x_{-i}^k) \geq \gamma - \varepsilon$. Since $\varepsilon > 0$ is arbitrary, it follows that $\limsup_k v_i(x_{-i}^k) \geq \gamma = \limsup_k v_i(\bar{x}_{-i}^k)$. Thus, u_i is i-upper semicontinuous. ∎

Theorem 5.11 allows us to state the existence of ε-equilibrium using only upper semicontinuity and quasiconcavity assumptions. A further particular case is obtained when players' payoff functions are piecewise polyhedral quasi-concave since, in this case, so are their value functions. Given a normal-form game $G = (X_i, u_i)_{i \in N}$ and $i \in N$, player i's payoff function is *piecewise polyhedral quasiconcave* if there is a finite cover of X_{-i} by compact convex subsets X_{-i}^l such that u_i is polyhedral quasiconcave on $X_i \times X_{-i}^l$ for all l.

Theorem 5.12   *If $G = (X_i, u_i)_{i \in N} \in \mathbb{G}_q$ is such that, for all $i \in N$, X_i is a polytope, u_i is upper semicontinuous and piecewise quasiconcave and v_i is polyhedral piecewise quasiconcave, then G has an ε-equilibrium for all $\varepsilon > 0$.*

In particular, if G is such that, for all $i \in N$, X_i is a polytope and u_i is upper semicontinuous and piecewise polyhedral quasiconcave, then G has an ε-equilibrium for all $\varepsilon > 0$.

5.5   Continuity of the Nash Equilibrium Correspondence

We return to the limit results of Section 5.3 and show how to obtain the upper hemicontinuity of the Nash equilibrium correspondence from them.

We focus on the stability of Nash equilibria with respect to changes in players' payoff functions. Fix a finite set of players $N = \{1, \ldots, n\}$ and compact metric spaces $\{X_i\}_{i \in N}$. Recall that the Nash equilibrium correspondence is $E : \mathbb{G}_s(X) \rightrightarrows X$ defined by

$$E(G) = \{x \in X : x \text{ is a Nash equilibrium of } G\}$$

for all $G \in \mathbb{G}_s(X)$.

Theorem 5.13   *Let $G = (X_i, u_i)_{i \in N} \in \mathbb{G}$ be sequentially better-reply secure. Then the equilibrium correspondence $E : \mathbb{G}_s(X) \rightrightarrows X$ is upper hemicontinuous and compact-valued at G.*

Proof.   Let $G = (X_i, u_i)_{i \in N} \in \mathbb{G}$ be sequentially better-reply secure. Let $\{G_k\}_{k=1}^\infty \subseteq \mathbb{G}_s(X)$ be such that $\lim_k d(G_k, G) = 0$ and $\{x_k\}_{k=1}^\infty \subseteq X$

be such that $x_k \in E(G_k)$ for all $k \in \mathbb{N}$. Since X is compact, we may assume that $\{x_k\}_{k=1}^{\infty}$ converges; let $x = \lim_k x_k$. It follows by Theorem 5.2 that $x \in E(G)$. Hence, by Theorem A.4, E is upper hemicontinuous at G. ∎

The upper hemicontinuity of the Nash equilibrium correspondence is a useful property for the existence of Nash equilibria. In fact, if a sequentially better-reply secure game G can be approximated with a sequence of games, each of whom have Nash equilibria, then the upper hemicontinuity of the Nash equilibrium correspondence implies that G also has a Nash equilibrium. This argument is illustrated with a result establishing the existence of mixed-strategy Nash equilibria in sequentially better-reply secure games.

Let $S \subseteq \mathbb{G}_s(X)$ be the set of all sequentially better-reply secure games $G \in \mathbb{G}_s(X)$ such that $E(G) \neq \emptyset$. As a consequence of Fort's Theorem A.9, we obtain that $E|_S$ is continuous except at a first category set.

Theorem 5.14 *Let $C = \{G \in S : E|_S$ is continuous at $G\}$. Then C^c is first category in S.*

The interpretation of Theorem 5.14 is that the set of games in S at which the restriction of the Nash equilibrium correspondence to S fails to be continuous is small. However, Theorem 5.14 does not exclude the possibility that its complement, i.e. the set of games in S at which the restriction of the Nash equilibrium correspondence to S is continuous, is also small.

Due to the above, we will focus on a smaller class of games where one will be able to make the case that equilibrium correspondence is generically continuous. Assume, in addition, that $X_1, \ldots, X_n$ are convex subsets of a vector space. Let S_1 be the set of all generalized payoff secure and sum-usc games $G \in \mathbb{G}_q(X)$. Moreover, let S_2 be the set of all weakly payoff secure and upper semicontinuous games $G \in \mathbb{G}_q(X)$.

Theorem 5.15 *The set $S_1 \cup S_2$ is closed in $\mathbb{G}_q(X)$.*

Proof. Note that it suffices to show that S_j is closed for all $j = 1, 2$.

We first consider S_1. Let $G \in \mathbb{G}_q(X)$ and $\{G_k\}_{k=1}^{\infty} \subseteq S_1$ be such that $\lim_k d(G_k, G) = 0$. Let $f = \sum_{i=1}^{n} u_i$ and $f_k = \sum_{i=1}^{n} u_i^k$ for all $k \in \mathbb{N}$. Note that $\lim_k \|f_k - f\| = 0$ since, for all $x \in X$, $|f_k(x) - f(x)| = |\sum_{i=1}^{n}(u_i^k(x) - u_i(x))| \leq \sum_{i=1}^{n} \|u_i^k - u_i\| \leq n\|u_k - u\|$. Furthermore, for all $\alpha \in \mathbb{R}$, $\{x \in X : f(x) \geq \alpha\} = \cap_{k=1}^{\infty}\{x \in X : f_k(x) \geq \alpha - \|f_k - f\|\}$. Hence, since G_k is sum-usc then $\{x \in X : f_k(x) \geq \alpha - \|f_k - f\|\}$ is closed for all $k \in \mathbb{N}$ and, thus, so is $\{x \in X : f(x) \geq \alpha\}$. In conclusion, G is sum-usc.

An analogous argument shows that $u_i(\cdot, x_{-i})$ is quasiconcave for all $i \in N$ and $x_{-i} \in X_{-i}$. In fact, for all $\alpha \in \mathbb{R}$, $i \in N$ and $x_{-i} \in X_{-i}$, $\{x_i \in X_i : u_i(x_i, x_{-i}) \geq \alpha\} = \cap_{k=1}^{\infty}\{x \in X : u_i^k(x_i, x_{-i}) \geq \alpha - \|u_i^k - u_i\|\}$ is convex since $\{x \in X : u_i^k(x_i, x_{-i}) \geq \alpha - \|u_i^k - u_i\|\}$ is convex for all $k \in \mathbb{N}$.

We next show that G is generalized payoff secure. Let $i \in N$, $x \in X$ and $\varepsilon > 0$. Let $k \in \mathbb{N}$ be such that $\|u_k - u\| < \varepsilon/4$. Since G_k is generalized payoff secure, then there exists $V \in N(x_{-i})$ and a well-behaved correspondence $\varphi_i : V \rightrightarrows X_i$ such that $u_i^k(z) \geq u_i^k(x) - \varepsilon/4$ for all $z \in \text{graph}(\varphi)$. Hence, for all $z \in \varphi_i$, $u_i(z) > u_i^k(z) - \varepsilon/4 > u_i^k(x) - \varepsilon/2 > u_i(x) - \varepsilon$. This shows that G is generalized payoff secure. Hence, together with the fact that G is sum-usc, we obtain that $G \in S_1$ and that S_1 is closed.

The proof that S_2 is closed is analogous to the one above and is, therefore, omitted. ∎

It follows from the above result that $S_1 \cup S_2$ is a complete metric space. Furthermore, for all $G \in S_1 \cup S_2$, we have that $E(G) \neq \emptyset$ by Theorems 3.2, 3.17 and 3.20. Hence, by combining Theorems A.10 and A.9, we obtain the generic continuity of $E|_{S_1 \cup S_2}$.

Theorem 5.16 *Let $C = \{G \in S_1 \cup S_2 : E|_{S_1 \cup S_2} \text{ is continuous at } G\}$. Then C is second category in $S_1 \cup S_2$.*

5.6 Strategic Approximation

We consider an alternative notion of robustness which puts an emphasis on finite action approximations. As we have seen in Theorem 5.5, approximations and robustness of equilibria to such approximations are useful to establish existence of equilibria; in particular, finite action approximations are convenient because the existence of a mixed strategy Nash equilibrium is immediate for the approximating games. Moreover, finite action approximations are particularly appealing when one regards the choice of infinite action spaces as a mathematical convenient way of modeling a situation which, in reality, has a large but finite action spaces. Whenever this is the case, the appropriateness of the infinite-action formalization is strengthened by the existence of at least one sequence of finite-action

approximations with the property that limit points of Nash equilibria of
the approximating games are Nash equilibria of the limit game.

We next consider one such notion. Given a normal-form game $G = (X_i, u_i)_{i \in N}$, recall that the set of mixed strategy Nash equilibria of G is $E(\bar{G})$, i.e. the set of pure strategy Nash equilibria of the mixed extension $\bar{G} = (M_i, u_i)_{i \in N}$ of G. A *strategic approximation of G* is a countable set of pure strategies $X^\infty = \prod_{i \in N} X_i^\infty$ of G such that, whenever, for all $i \in N$, $\{X_i^k\}_{k=1}^\infty$ is such that X_i^k is finite, $X_i^k \subseteq X_i^{k+1}$ for all $k \in \mathbb{N}$ and $X_i^\infty \subseteq \cup_{k=1}^\infty X_i^k$, then $\mathrm{Li}(E(\bar{G}_k)) \subseteq E(\bar{G})$, where $G_k = (X_i^k, u_i)_{i \in N}$ for all $k \in \mathbb{N}$.

The notion of a strategic approximation is related to the notion of weak sequential better-reply security introduced in Section 5.3. In fact, arguing as in the proof of Theorem 5.2, we have that G is weakly sequential better-reply secure with respect to $\{G_k\}_{k=1}^\infty$ for all sequences $\{G_k\}_{k=1}^\infty$ as in the definition of a strategic approximation of G. Note, however, that games that have a strategic approximation may fail to be weakly sequential better-reply secure, simply because the strategy spaces in the approximations must eventually include X^∞. Nevertheless, because any such approximation that eventually includes X^∞ is admissible, there is robustness in the choice of the finite-action approximations. Thus, strategic approximation is a notion of robust approximation of an infinite-action game by finite-action games whose requirement is strengthened by only allowing approximations that eventually include strategies that are particularly relevant, for instance, because they may have particular strategic significance to the players.

The point can be illustrated by the game in Example 4.13. Recall that $N = \{1, 2\}$, $X_1 = X_2 = [0, 1]$,

$$u_1(x_1, x_2) = \begin{cases} 1 & \text{if } x_1 = 1, \\ 0 & \text{if } x_1 < 1 \text{ and } x_2 = 1/2, \\ 2x_2 & \text{if } x_1 < 1 \text{ and } x_2 < 1/2, \\ 2(1 - x_2) & \text{if } x_1 < 1 \text{ and } x_2 > 1/2, \end{cases}$$

and $u_2(x_1, x_2) = u_1(x_2, x_1)$ for all $(x_1, x_2) \in X$.

Suppose that, for all $k \in \mathbb{N}$ and $i \in N$, we let $X_i^k = \{j/k : j = 0, \ldots, k - 1\}$. Thus, $X_i^k \subseteq X_i^{k+1}$, $\lim_k \delta_i(X_i^k, X_i) = 0$ and $1 \notin X_i^k$ for all $i \in N$ and $k \in \mathbb{N}$. The latter fact, i.e., $1 \notin X_i^k$ for all $i \in N$ and $k \in \mathbb{N}$, will be responsible for the failure of $\mathrm{Li}(E(\bar{G}_k)) \subseteq E(\bar{G})$. Indeed, letting j_k be such that $j_k = \max\{j \in \{0, \ldots, k - 1\} : j_k/k < 1/2\}$ and $x_k =$

$(j_k/k, j_k/k)$ for all $k \in \mathbb{N}$, it follows that $x_k \in E(\bar{G}_k)$, $\lim_k x_k = (1/2, 1/2)$ and $(1/2, 1/2) \notin E(\bar{G})$.

There is a clear sense in which $x_i = 1$ has particular strategic significance for player i in the above example: in fact, such strategy is strictly dominant. Once such strategy is eventually included in the sequence, say $(1, 1) \in X_k$ for all $k \geq K$, then $m_k \in E(\bar{G}_k)$ implies that $m_k = (1, 1) \in X$. Since $E(\bar{G}) = \{(1, 1)\}$, then it follows that $\{(1, 1)\}$ is a strategic approximation of G.

An advantage of the notion of strategic approximation is that, by dropping the requirement that all finite-action approximations have to be considered, it holds for games, such as the above, that fail to be sequentially better-reply secure.

We next consider two conditions that are sufficient for the existence of strategic approximations. We say that G has the *finite deviation property* if whenever $m \in M$ is not a mixed strategy Nash equilibrium of G, there exists a neighborhood U of m and a finite subset D of M such that, for every $m' \in U$, there exists $i \in N$ and $\hat{m} \in D$ such that $u_i(\hat{m}_i, m'_{-i}) > u_i(m')$. Furthermore, G has the *finite-support finite deviation property* if the members of D can always be chosen to have finite supports.

Before we establish the existence of strategic approximations, we note that the finite deviation property is implied by C-security. The notion of C-security in quasiconcave games is obtained by requiring players to secure payoffs using a single strategy instead of using a well-behaved correspondence as in generalized C-security. Formally, given a normal-form game $G = (X_i, u_i)_{i \in N}$, we say that G is *C-secure* if, whenever $x \notin E(G)$, there exists an open neighborhood U of x, $\bar{x} \in X$ and $\alpha \in \mathbb{R}^n$ such that

(a) $u_i(\bar{x}_i, x'_{-i}) \geq \alpha_i$ for all $i \in N$ and $x' \in U$, and
(b) For all $x' \in U$, there exists $i \in N$ such that $u_i(x') < \alpha_i$.

Furthermore, in the particular case of the mixed extension $\bar{G} = (M_i, u_i)_{i \in N}$ of a normal-form game $G = (X_i, u_i)_{i \in N}$, we say that $\bar{G}$ is *finite-support C-secure* if the securing strategy can be chosen to have finite support. More precisely, $\bar{G}$ is finite-support C-secure if, whenever $m \notin E(\bar{G})$, there exists an open neighborhood U of m, $\bar{m} \in M$ with finite support and $\alpha \in \mathbb{R}^n$ such that, for all $m' \in U$, $u_i(\bar{m}_i, m'_{-i}) \geq \alpha_i$ for all $i \in N$ and $u_i(m') < \alpha_i$ for some $i \in N$.

Theorem 5.17   *Let $G = (X_i, u_i)_{i \in N} \in \mathbb{G}$. Then:*

1. *If $\bar{G}$ is C-secure, then G has the finite deviation property.*

2. *If $\bar{G}$ is finite-support C-secure, then G has the finite-support finite deviation property.*

   Proof.   Suppose that $\bar{G}$ is C-secure and let $m \notin E(\bar{G})$. Then there exists an open neighborhood U of m, $\bar{m} \in M$ and $\alpha \in \mathbb{R}^n$ such that, for all $m' \in U$, $u_i(\bar{m}_i, m'_{-i}) \geq \alpha_i$ for all $i \in N$ and $u_i(m') < \alpha_i$ for some $i \in N$. Hence, for all $m' \in U$, there exists $i \in N$ such that $u_i(\bar{m}_i, m'_{-i}) \geq \alpha_i > u_i(m')$. This shows that G has the lower deviation property.

   Finally, note that an analogous argument shows that G has the finite-support finite deviation property if $\bar{G}$ is finite-support C-secure.   ∎

   We next show that the finite deviation property suffices for the existence of a strategic approximation of the mixed extension of a game. Furthermore, the finite-support finite deviation property is sufficient for the existence of a strategic approximation of a game.

Theorem 5.18   *Let $G = (X_i, u_i)_{i \in N} \in \mathbb{G}$. Then:*

1. *If G has the finite deviation property, then $\bar{G}$ has a strategic approximation.*
2. *If G has the finite-support finite deviation property, then G has a strategic approximation.*

   Note that the definition of a strategic approximation considers mixed strategy equilibria. Thus, in particular, a strategic approximation of a mixed extension involves mixed strategy equilibria of the mixed extension. The mixed extension of the mixed extension of a normal-form game $G = (X_i, u_i)_{i \in N}$ is a normal-form game with the same set of players N and where each player's strategy space is the space of Borel probability measures on M_i. We denote such space by $\Delta(M_i)$ and let $\Delta(M) = \prod_{i \in N} \Delta(M_i)$. Player i' payoff function is also denoted by u_i and is the extension of $u_i : M \to \mathbb{R}$ to $\Delta(M)$, i.e. $u_i(\mu) = \int_M u_i(m) d\mu(m)$ for all $\mu \in \Delta(M)$.

   The following result establishes useful results concerning the relationship between any strategy $\mu \in \Delta(M)$ and its *distribution on X*, $\bar{m} \in M$, defined by $\bar{m}(B) = \int_M m(B) d\mu(m)$ for every Borel subset B of X.

Lemma 5.19   *Let $\mu \in \Delta(M)$ and $\bar{m} \in M$ be its distribution on X. Then:*

1. *$\bar{m}$ is well-defined.*
2. *$u_i(\mu) = u_i(\bar{m})$ for all $i \in N$.*

3. *If $\{\mu_k\}_{k=1}^\infty \subseteq \Delta(M)$ and $\{\bar{m}_k\}_{k=1}^\infty \subseteq M$ are such that $\bar{m}_k$ is the distribution of μ_k on X for all $k \in \mathbb{N}$ and $\lim_k \mu_k = \mu$, then $\lim_k \bar{m}_k = \bar{m}$.*

Proof. Part 1 follows because the function $m \mapsto m(B)$ is measurable for every Borel subset B of X by Theorem A.22.

We turn to part 2. Given a Borel subset B of X, let χ_B be its characteristic function. Then, by definition, we have that $\int_X \chi_B(x)\,\mathrm{d}\bar{m}(x) = \bar{m}(B) = \int_M m(B)\mathrm{d}\mu(m) = \int_M \int_X \chi_B(x)\mathrm{d}m(x)\mathrm{d}\mu(m)$. Hence, it follows that for all bounded Borel measurable functions $f : X \to \mathbb{R}$, $\int_X f(x)\mathrm{d}\bar{m}(x) = \int_M \int_X f(x)\mathrm{d}m(x)\mathrm{d}\mu(m) = \int_M f(m)\mathrm{d}\mu(m)$. In particular, $u_i(\bar{m}) = \int_X u_i(x)\mathrm{d}\bar{m}(x) = \int_M u_i(m)\mathrm{d}\mu(m) = u_i(\mu)$ for all $i \in N$.

Regarding part 3, let $f : X \to \mathbb{R}$ be continuous and bounded. Note that $m \mapsto \int_X f\mathrm{d}m$ is continuous by Theorem A.23 and, therefore, $\lim_k \int_M \int_X f\mathrm{d}m\mathrm{d}\mu_k = \int_M \int_X f\mathrm{d}m\mathrm{d}\mu$. Hence, by part 2,

$$\lim_k \int_X f\mathrm{d}\bar{m}_k = \lim_k \int_M \int_X f\mathrm{d}m\mathrm{d}\mu_k = \int_M \int_X f\mathrm{d}m\mathrm{d}\mu = \int_X f\mathrm{d}\bar{m}.$$

This completes the proof. ∎

We turn to the proof of Theorem 5.18.

Proof of Theorem 5.18. Let $G = (X_i, u_i)_{i \in N} \in \mathbb{G}$ and suppose that G has the finite deviation property. For convenience, let $O = E(\bar{G})^c$. Thus, for all $m \in O$, there exists an open neighborhood U_m of m and a finite set D_m satisfying the definition of the finite deviation property. Since $\{U_m\}_{m \in O}$ is an open cover of O, O is open and X is separable (being compact), then there exists a countable subcollection $\{(U_j, D_j)\}_{j=1}^\infty$ such that $O = \cup_{j=1}^\infty U_j$. For all $i \in N$ and $k \in \mathbb{N}$, let D_i^j be the projection of D_j onto M_i and define $M_i^\infty = \cup_{k=1}^\infty D_i^j$. We clearly have that $M^\infty = \prod_{i \in N} M_i^\infty$ is countable. We next show that M^∞ is a strategic approximation of $\bar{G}$.

Let, for all $i \in N$, $\{M_i^k\}_{k=1}^\infty$ be an increasing sequence of finite subsets of M_i whose union contains M_i^∞. Furthermore, let $\{\mu_k\}_{k=1}^\infty$ be such that μ_k is a mixed strategy equilibrium of $G_k = (M_i^k, u_i)_{i \in N}$ for all $k \in \mathbb{N}$ and let $\mu^* \in \Delta(M)$ be such that $\lim_k \mu_k = \mu^*$.

Let m_k be the distribution of μ_k on X for all $k \in \mathbb{N}$ and m^* be the distribution of μ^* on X. We first claim that m^* is a mixed strategy Nash equilibrium of G.

Suppose, in order to reach a contradiction, that $m^* \in O$. Then, for some $j \in \mathbb{N}$, $m^* \in U_j$. Since $\lim_k \mu_k = \mu^*$, then $\lim_k m_k = m^*$ by Lemma 5.19.

Furthermore, $D_i^j \subseteq \cup_{k=1}^\infty M_i^k$ for all $i \in N$. Hence, there is $K \in \mathbb{N}$ such that $m_k \in U_j$ and $D_i^j \subseteq M_i^k$ for all $i \in N$ and $k \geq K$.

Thus, for all $k \geq K$, since $m_k \in U_j$, there exists $i \in N$ and $\hat{m}_i \in D_i^j \subseteq M_i^k$ such that $u_i(\hat{m}_i, m_{-i}^k) > u_i(m_k)$. It follows by Lemma 5.19 that $u_i(\hat{m}_i, \mu_{-i}^k) = u_i(\hat{m}_i, m_{-i}^k) > u_i(m_k) = u_i(\mu_k)$. Since $\hat{m}_i \in M_i^k$, this contradicts the assumption that μ_k is a mixed strategy Nash equilibrium of G_k.

The above contradiction shows that m^* is a mixed strategy Nash equilibrium of G. This together with Lemma 5.19 implies that, for all $i \in N$ and $m_i \in M_i$, $u_i(\mu^*) = u_i(m^*) \geq u_i(m_i, m_{-i}^*) = u_i(m_i, \mu_{-i}^*)$. Thus, μ^* is a mixed strategy Nash equilibrium of $\bar{G}$ as desired.

We next turn to the second part of Theorem 5.18. Suppose that G has the finite-support finite deviation property and let M^∞ be defined as above. Since G has the finite-support finite deviation property, we may assume that each $m \in M^\infty$ has a finite support. For all $i \in N$ and $m_i \in M_i$, let $\mathrm{supp}(m_i)$ denote the support of m_i and define $X_i^\infty = \cup_{m_i \in M_i^\infty} \mathrm{supp}(m_i)$ and $X^\infty = \prod_{i \in N} X_i^\infty$. Note that X^∞ is countable because M^∞ is countable. We next show that X^∞ is a strategic approximation of G.

Let, for all $i \in N$, $\{X_i^k\}_{k=1}^\infty$ be an increasing sequence of finite subsets of X_i whose union contains X_i^∞. Furthermore, let $\{m_k\}_{k=1}^\infty$ be such that m_k is a mixed strategy equilibrium of $G_k = (X_i^k, u_i)_{i \in N}$ for all $k \in \mathbb{N}$ and let $m^* \in M$ be such that $\lim_k m_k = m^*$.

Suppose, in order to reach a contradiction, that $m^* \in O$. Then, for some $j \in \mathbb{N}$, $m^* \in U_j$. Furthermore, $\cup_{m_i \in D_i^j} \mathrm{supp}(m_i) \subseteq \cup_{k=1}^\infty X_i^k$ for all $i \in N$. Hence, there is $K \in \mathbb{N}$ such that $m_k \in U_j$ and $\cup_{m_i \in D_i^j} \mathrm{supp}(m_i) \subseteq X_i^k$ for all $i \in N$ and $k \geq K$.

Thus, for all $k \geq K$, since $m_k \in U_j$, there exists $i \in N$ and $\hat{m}_i \in D_i^j$ such that $u_i(\hat{m}_i, m_{-i}^k) > u_i(m_k)$. This, together with $\mathrm{supp}(\hat{m}_i) \in X_i^k$, contradict the assumption that m_k is a mixed strategy Nash equilibrium of G_k. Thus, it follows that m^* is a mixed strategy of G as desired. ∎

In the particular case where a game $G = (X_i, u_i)_{i \in N}$ is such that $\bar{G}$ is better-reply secure, we have that $\bar{G}$ is C-secure (by an argument analogous to the one used in the proof of Theorem 4.11) and, therefore, $\bar{G}$ has a strategic approximation. The next result provides a sufficient condition for games whose mixed extension is better-reply secure to have a strategic approximation.

Theorem 5.20   *Let $G = (X_i, u_i)_{i \in N} \in \mathbb{G}$. If $\bar{G}$ is better-reply secure and, for all $i \in N$ and $x_{-i} \in X_{-i}$, the complement of the set $\{x_i \in X_i : u_i$ is continuous at $(x_i, x_{-i})\}$ is countable, then G has the finite-support finite deviation property. Consequently, G has a strategic approximation.*

The following lemma is used in the proof of the above result. Recall that $\bar{u}_i(m) = \sup_{U \in N(m_{-i})} \inf_{m'_{-i} \in U} u_i(m_i, m'_{-i})$ for all $i \in N$ and $m \in M$, where $N(m_{-i})$ is the set of open neighborhoods of m_{-i}.

Lemma 5.21   *Let $G = (X_i, u_i)_{i \in N} \in \mathbb{G}$, $i \in N$, $\nu_i, \lambda_i \in M_i$, $m_{-i} \in M_{-i}$ and $a \in [0, 1]$. Then the following holds:*

1. $\bar{u}_i(a\nu_i + (1 - a)\lambda_i, m_{-i}) \geq a\bar{u}_i(\nu_i, m_{-i}) + (1 - a)\bar{u}_i(\lambda_i, m_{-i})$.
2. *If u_i is continuous at (ν_i, m_{-i}), then $\bar{u}_i(a\nu_i + (1 - a)\lambda_i, m_{-i}) = a\bar{u}_i(\nu_i, m_{-i}) + (1 - a)\bar{u}_i(\lambda_i, m_{-i})$.*

Proof.   Given $\varepsilon > 0$, let $U_1, U_2 \in N(m_{-i})$ be such that $\inf_{m'_{-i} \in U_1} u_i(\nu_i, m'_{-i}) > \bar{u}_i(\nu_i, m_{-i}) - \varepsilon$ and $\inf_{m'_{-i} \in U_2} u_i(\lambda_i, m'_{-i}) > \bar{u}_i(\lambda_i, m_{-i}) - \varepsilon$. Then, letting $U = U_1 \cap U_2$ and $m'_{-i} \in U$, we have that $u_i(a\nu_i + (1 - a)\lambda_i, m'_{-i}) = au_i(\nu_i, m'_{-i}) + (1 - a)u_i(\lambda_i, m'_{-i}) > a\bar{u}_i(\nu_i, m_{-i}) + (1 - a)\bar{u}_i(\lambda_i, m_{-i}) - \varepsilon$. Thus, $\bar{u}_i(a\nu_i + (1 - a)\lambda_i, m_{-i}) \geq a\bar{u}_i(\nu_i, m_{-i}) + (1 - a)\bar{u}_i(\lambda_i, m_{-i}) - \varepsilon$ and part 1 follows by noting that $\varepsilon > 0$ is arbitrary.

We turn to part 2. Note that part 1 already implies that $\bar{u}_i(a\nu_i + (1 - a)\lambda_i, m_{-i}) \geq a\bar{u}_i(\nu_i, m_{-i}) + (1 - a)\bar{u}_i(\lambda_i, m_{-i})$. Given $\varepsilon > 0$, the definition of $\bar{u}_i$ and the continuity of u_i at (ν_i, m_{-i}) implies that there is $U \in N(m_{-i})$ such that $\inf_{m'_{-i} \in U} u_i(a\nu_i + (1 - a)\lambda_i, m'_{-i}) > \bar{u}_i(a\nu_i + (1 - a)\lambda_i, m_{-i}) - \varepsilon$ and $|u_i(\nu_i, m'_{-i}) - u_i(\nu_i, m_{-i})| < \varepsilon$ for all $m'_{-i} \in U$.

For all $m'_{-i} \in U$, we have that $au_i(\nu_i, m'_{-i}) + (1 - a)u_i(\lambda_i, m'_{-i}) = u_i(a\nu_i + (1 - a)\lambda_i, m'_{-i}) > \bar{u}_i(a\nu_i + (1 - a)\lambda_i, m_{-i}) - \varepsilon$. Hence,

$$(1 - a)u_i(\lambda_i, m'_{-i}) > \bar{u}_i(a\nu_i + (1 - a)\lambda_i, m_{-i}) - \varepsilon - au_i(\nu_i, m'_{-i})$$

$$> \bar{u}_i(a\nu_i + (1 - a)\lambda_i, m_{-i}) - au_i(\nu_i, m_{-i}) - 2\varepsilon.$$

Therefore, $(1 - a)\bar{u}_i(\lambda_i, m_{-i}) \geq \bar{u}_i(a\nu_i + (1 - a)\lambda_i, m_{-i}) - au_i(\nu_i, m_{-i}) - 2\varepsilon$ and so

$$a\bar{u}_i(\nu_i, m_{-i}) + (1 - a)\bar{u}_i(\lambda_i, m_{-i}) =$$

$$au_i(\nu_i, m_{-i}) + (1 - a)\bar{u}_i(\lambda_i, m_{-i}) \geq \bar{u}_i(a\nu_i + (1 - a)\lambda_i, m_{-i}) - 2\varepsilon.$$

The conclusion follows by noting that $\varepsilon > 0$ is arbitrary. ∎

We now proceed to the proof of Theorem 5.20.

Proof of Theorem 5.20.   Let $G = (X_i, u_i)_{i \in N}$ satisfy the assumptions of the theorem. Since $\bar{G}$ is better-reply secure, by an argument analogous to that in Theorem 4.11, for all $m \notin E(\bar{G})$ there exist $\alpha \in \mathbb{R}^n$, $\varepsilon > 0$, an open neighborhood U of m and $\hat{m} \in M$ such that, for all $m' \in U$, $u_i(\hat{m}_i, m'_{-i}) \geq \alpha_i + \varepsilon$ for all $i \in N$, and $u_i(m') < \alpha_i - \varepsilon$ for some $i \in N$.

Hence, a sufficient condition for $\bar{G}$ to be finite-support C-secure and, therefore by Theorem 5.17, for G to have the finite-support finite deviation property is the following: For all $i \in N$, $m \in M$ and $\varepsilon > 0$, there exists $\bar{m}_i \in M_i$ such that $\text{supp}(\bar{m}_i)$ is finite and $\bar{u}_i(\bar{m}_i, m_{-i}) \geq \bar{u}_i(m) - \varepsilon$.

We establish the above property in the remainder of this proof. Let $i \in N$, $m \in M$ and $\varepsilon > 0$. Denote $A = \{x_i \in X_i : m_i(x_i) > 0\}$ and note that A is countable. Indeed, $A = \cup_{k=1}^{\infty} \{x_i \in X_i : m_i(x_i) \geq 1/k\}$ and, since m_i is a probability measure, $\{x_i \in X_i : m_i(x_i) \geq 1/k\}$ has at most k elements.

Define the following two measures λ'_i and ν'_i on X_i. For all Borel measurable subsets B of X_i, let $\lambda'_i(B) = m_i(A \cap B)$ and $\nu'_i(B) = m_i(B) - \lambda'_i(B)$. Clearly, we have that λ'_i and ν'_i are indeed measures on X_i and that $\nu'_i(\{x_i\}) = 0$ for all $x_i \in X_i$.

Note that, in general, neither ν'_i nor λ'_i are probability measures on X_i. In the case where $\nu'_i(X_i) > 0$ and $\lambda'_i(X_i) > 0$, define $\nu_i = \nu'_i/\nu'_i(X_i)$ and $\lambda_i = \lambda'_i/\lambda'_i(X_i)$. The goal is to approximate both ν_i and λ_i by corresponding probability measures with finite support. Before we do that, we note that we cannot have $\nu'_i(X_i) = \lambda'_i(X_i) = 0$ and that in the case where one of them is zero, say $\nu'_i(X_i) = 0$, then $\lambda'_i = \lambda_i$ and we only approximate λ_i by a finite support probability measure.

Let $a = \nu'_i(X_i)$ and note that $1 - a = m_i(X_i) - \nu'_i(X_i) = \lambda'_i(X_i)$. Furthermore, $a\nu_i + (1 - a)\lambda_i = m_i$.

Since A is countable, let $A = \{x_i^j\}_{j=1}^{\infty}$. For all $k \in \mathbb{N}$, let λ_i^k be defined as follows: $\lambda_i^k(x_i^1) = \lambda_i(x_i^1) + \sum_{j=k+1}^{\infty} \lambda_i(x_i^j)$, $\lambda_i^k(x_i^j) = \lambda_i(x_i^j)$ for all $2 \leq j \leq k$ and $\lambda_i^k(x_i^j) = 0$ for all $j > k$. Clearly, $\text{supp}(\lambda_i^k) = \{x_i^1, \ldots, x_i^k\}$.

Furthermore, for all $\gamma > 0$, there is $K \in \mathbb{N}$ such that $|u_i(\lambda_i^k, m'_{-i}) - u_i(\lambda_i, m'_{-i})| < \gamma$ for all $k \geq K$ and $m'_{-i} \in M_{-i}$. Indeed, letting $B > 0$ be such that $\|u\| \leq B$ and $K \in \mathbb{N}$ be such that $2B\sum_{j=k+1}^{\infty} \lambda_i(x_j) < \gamma$ for all

$k \geq K$, we have that, for all $k \geq K$ and $m'_{-i} \in M_{-i}$,

$$|u_i(\lambda_i^k, m'_{-i}) - u_i(\lambda_i, m'_{-i})| = \left| \sum_{j=k+1}^{\infty} \lambda_i(x_j)[u_i(x_i^1, m'_{-i}) - u_i(x_i^j, m'_{-i})] \right|$$

$$\leq 2B \sum_{j=k+1}^{\infty} \lambda_i(x_j) < \gamma.$$

As a result of the above, we obtain that

$$\bar{u}_i(\lambda_i^k, m_{-i}) \geq \bar{u}_i(\lambda_i, m_{-i}) - \varepsilon$$

for all k sufficiently large. Indeed, there exists $U \in N(m_{-i})$ such that $\inf_{m'_{-i} \in U} u_i(\lambda_i, m'_{-i}) > \bar{u}_i(\lambda_i, m_{-i}) - \varepsilon/2$. Furthermore, there exists $K \in \mathbb{N}$ such that $|u_i(\lambda_i^k, m'_{-i}) - u_i(\lambda_i, m'_{-i})| < \varepsilon/2$ for all $k \geq K$ and $m'_{-i} \in M_{-i}$. Hence, given $k \geq K$, for all $m'_{-i} \in U$, it follows that $u_i(\lambda_i^k, m'_{-i}) > u_i(\lambda_i, m'_{-i}) - \varepsilon/2 > \bar{u}_i(\lambda_i, m_{-i}) - \varepsilon$ and so $\bar{u}_i(\lambda_i^k, m_{-i}) \geq \bar{u}_i(\lambda_i, m_{-i}) - \varepsilon$.

Turning to ν_i, note that $\nu_i(\{x_i\}) = 0$ for all $x_i \in X_i$ together with the assumption that, for all $x_{-i} \in X_{-i}$, the complement of the set $\{x_i \in X_i : u_i \text{ is continuous at } (x_i, x_{-i})\}$ is countable implies that

$$\nu_i(\{x_i \in X_i : u_i \text{ is discontinuous at } (x_i, x_{-i})\}) = 0 \text{ for all } x_{-i} \in X_{-i}.$$

Hence,

$$(\nu_i, m_{-i})(\{x \in X : u_i \text{ is discontinuous at } x\})$$

$$= \int_{X_{-i}} \nu_i(\{x_i \in X_i : u_i \text{ is discontinuous at } (x_i, x_{-i})\}) dm_{-i}(x_{-i}) = 0.$$

It then follows by Theorem A.23 that u_i is continuous at (ν_i, m_{-i}).

By Theorem A.20 there exists a sequence of measures with finite support $\{\nu_i^k\}_{k=1}^{\infty} \subseteq M_i$ such that $\lim_k \nu_i^k = \nu_i$. The continuity of u_i at (ν_i, m_{-i}) implies that $\bar{u}_i(\nu_i, m_{-i}) = u_i(\nu_i, m_{-i})$ and that

$$\bar{u}_i(\nu_i^k, m_{-i}) > u_i(\nu_i, m_{-i}) - \varepsilon$$

for all sufficiently large k. To see the latter, let $O \in N(\nu_i, m_{-i})$ be such that $|u_i(m') - u_i(\nu_i, m_{-i})| < \varepsilon$ for all $m' \in O$ and let $V \in N(\nu_i)$ and $U \in N(m_{-i})$ be such that $V \times U \subseteq O$. Then, for all k such that $\nu_i^k \in V$

(hence, for all k sufficiently large), and for all $m'_{-i} \in U$, $u_i(\nu_i^k, m'_{-i}) > u_i(\nu_i, m_{-i}) - \varepsilon$ implies that $\inf_{m'_{-i} \in U} u_i(\nu_i^k, m'_{-i}) \geq u_i(\nu_i, m_{-i}) - \varepsilon$. Thus, $\bar{u}_i(\nu_i^k, m_{-i}) \geq u_i(\nu_i, m_{-i}) - \varepsilon$ for all sufficiently large k.

We finally define $\bar{m}_i$. Recall that $a = \nu_i'(X_i)$, $1 - a = \lambda_i'(X_i)$ and $a\nu_i + (1-a)\lambda_i = m_i$. Let $k \in \mathbb{N}$ be such that $\bar{u}_i(\nu_i^k, m_{-i}) \geq u_i(\nu_i, m_{-i}) - \varepsilon$ and $\bar{u}_i(\lambda_i^k, m_{-i}) \geq \bar{u}_i(\lambda_i, m_{-i}) - \varepsilon$ and define $\bar{m}_i = a\nu_i^k + (1-a)\lambda_i^k$.

Clearly, $\mathrm{supp}(\bar{m}_i)$ is finite. Furthermore, since u_i is continuous at (ν_i, m_{-i}), it follows by Lemma 5.21 that

$$
\begin{aligned}
\bar{u}_i(\bar{m}_i, m_{-i}) &= \bar{u}_i(a\nu_i^k + (1-a)\lambda_i^k, m_{-i}) \\
&\geq a\bar{u}_i(\nu_i^k, m_{-i}) + (1-a)\bar{u}_i(\lambda_i^k, m_{-i}) \\
&\geq au_i(\nu_i, m_{-i}) + (1-a)\bar{u}_i(\lambda_i, m_{-i}) - \varepsilon \\
&= a\bar{u}_i(\nu_i, m_{-i}) + (1-a)\bar{u}_i(\lambda_i, m_{-i}) - \varepsilon \\
&= \bar{u}_i(a\nu_i + (1-a)\lambda_i, m_{-i}) - \varepsilon \\
&= \bar{u}_i(m) - \varepsilon,
\end{aligned}
$$

as desired. This completes the proof. ∎

5.7   References

The material in Sections 5.2 and 5.3 is based on Carbonell-Nicolau and McLean (2011). In particular, the notions of sequential better-reply security and very weak better-reply security are due to Carbonell-Nicolau and McLean (2011).

Strong variational convergence and strong function approximation were introduced in Bagh (2010) and Castro (2011), respectively, in a weaker form. The notion of multi-hypoconvergence is taken from Lucchetti and Patrone (1986).

Theorem 5.7 is due to Carmona (2011c), who generalizes a result due to Reny (1996) and Prokopovych (2011).

The notion of piecewise quasiconcavity was introduced by Radzik (1991) under the name strong quasiconcavity and generalized by Ziad (1997). This latter paper introduced the concept of i-upper semicontinuity. Theorems 5.8, 5.11 and 5.12, as well as Lemmas 5.9 and 5.10, are due to Carmona (2010).

Section 5.5 is based on Carbonell-Nicolau (2010). In particular, Theorem 5.15 is due to Carbonell-Nicolau (2010).

Finally, the notion of strategic approximation, as well as the results in Section 5.6, are due to Reny (2011b).

Chapter 6

Games With an Endogenous Sharing Rule

This chapter presents an alternative way of addressing the existence of equilibrium. This approach is particularly useful in situations where there is some flexibility in defining players' payoff functions at strategies where these would naturally be discontinuous. The following example, due to Simon and Zame (1990), illustrates.

Let $N = \{1, 2\}$, $X_1 = [0, 3]$, $X_2 = [3, 4]$ and, for all $\alpha \in \mathbb{R}^2$ and $x \in X$,

$$
u_1(x) = \begin{cases} \dfrac{x_1 + x_2}{2} & \text{if } x_1 < x_2, \\ \alpha_1 & \text{if } x_1 = x_2, \end{cases}
$$

$u_2(x) = 4 - u_1(x)$ and $\alpha_2 = 4 - \alpha_1$. An interpretation of this game is that the players are psychologists who face a market where costumers are uniformly distributed over the interval $[0, 4]$. Psychologist 1 can only locate himself in California, represented by $[0, 3]$, whereas psychologist 2 can only locate herself in Oregon, which is represented by $[3, 4]$. Customers then choose the closest psychologist.

When psychologists locate in different locations (i.e., $x_i \neq 3$ for some $i \in N$), then psychologist 1 attracts costumers in $[0, (x_1 + x_2)/2]$ and psychologist 2 those in $[(x_1 + x_2)/2, 4]$. In this case, there is no ambiguity in the specification of payoff since those costumers who are indifferent between the two are those located at $(x_1 + x_2)/2$ and have, therefore, measure zero. However, this is no longer the case when they locate at the same location (i.e., $x_1 = x_2 = 3$). Indeed, in this case all costumers are indifferent between the two psychologists.

How should the market be slitted between the two psychologists, i.e., how should we choose α? Instead of specifying α a priori, we can look for those α that are part of an equilibrium. More precisely, the goal is to find pairs (x, α) such that x is a Nash equilibrium when $u(3,3) = \alpha$.

In the above game, it is clear that both players want to be as close as possible to the point 3 (the border between California and Oregon). More precisely, for all $x_2 \in X_2$, $u_1(x_1, x_2) > u_1(x_1', x_2)$ whenever $3 > x_1 > x_1' \geq 0$ and similarly for player 2. Hence, for each $i \in N$, player i's best-reply correspondence is either empty-valued or equal to $\{3\}$. Thus, a Nash equilibrium exists only if the latter holds for both players. This requires $\alpha_1 \geq (3 + x_2)/2$ and $\alpha_2 \geq (x_1 + 3)/2$, which is equivalent to $x_2 \leq 2\alpha_1 - 3$ and $x_1 \geq 5 - 2\alpha_2$. Since $x_1 \leq 3$ and $x_2 \geq 3$, such inequalities hold only if $\alpha_1 \geq 3$ and $\alpha_2 \geq 1$. This, together with $u_1 + u_2 = 4$, implies that $\alpha = (3, 1)$. In words, an equilibrium way of splitting the market when both psychologists locate in the border between California and Oregon is to assign those costumers in California to the psychologist from California and those in Oregon to the one from Oregon. Furthermore, the only equilibrium way of splitting the market in this case is such that the psychologist from California has 3/4 of the market share.

The above example can be understood as one where the way of sharing payoffs is endogenous. In the next section we present formal definitions of games with an endogenous sharing rule and of solutions for such games. We then establish an existence of result for games with an endogenous sharing rule. Such existence result will be derived from a limit result.

6.1   Existence and Stability of Solutions

A *game with an endogenous sharing rule* $\Gamma = (X_1, \ldots, X_n, Q)$ consists of a finite set of players $N = \{1, \ldots, n\}$, a compact metric strategy space X_i for each player $i \in N$ and a well-behaved payoff correspondence $Q : X \rightrightarrows \mathbb{R}^n$ (i.e., Q is upper hemicontinuous with nonempty, convex, compact values). A *solution* for Γ is a pair (q, α) such that q is a measurable selection of Q and $\alpha = (\alpha_1, \ldots, \alpha_n)$ is a mixed strategy Nash equilibrium of the normal-form game $G_q = (X_i, q_i)_{i \in N}$.

The following is a limit result for games with an endogenous sharing rules in the following sense: if $\{(q_k, \alpha_k)\}_{k=1}^{\infty}$ is a sequence of solutions of a game with an endogenous sharing rule Γ and $\alpha = \lim_k \alpha_k$, then α is part of a solution of Γ. An alternative interpretation arises when we focus on the sequence of normal-form games $G_k = (X_i, q_i^k)_{i \in N}$. Comparing Theorem 6.1

with the limit results presented in Chapter 5, we have that the former is a limit result without the specification of a limit game and, therefore, without any assumptions (such as sequential better-reply security) on such limit game. In contrast, the limit game is derived endogenously.

Theorem 6.1 *Let* $\Gamma = (X_1, \ldots, X_n, Q)$ *be a game with an endogenous sharing rule and* $\{G_k\}_{k=1}^{\infty}$ *be such that* $G_k = (X_i^k, u_i^k)_{i \in N}$ *for all* $k \in \mathbb{N}$.
Suppose that

1. X_i^k *is a Borel measurable subset of* X_i *and* $\mathrm{Li}(X_i^k) = X_i$ *for all* $k \in \mathbb{N}$ *and* $i \in N$, *and*
2. u_k *is measurable and* $u_k(x) \in Q(x)$ *for all* $x \in X_k$ *and* $k \in \mathbb{N}$.

Then, the following holds: If $\alpha \in M(X)$ *and* $\{\alpha_k\}_{k=1}^{\infty} \subseteq M(X)$ *are such that* $\lim_k \alpha_k = \alpha$ *and* α_k *is a mixed strategy Nash equilibrium of* G_k *for all* $k \in \mathbb{N}$, *then there exists a measurable selection* u *of* Q *such that* (α, u) *is a solution of* Γ *and* $\lim_k u_k(\alpha_k) = u(\alpha)$.

We turn to the proof of Theorem 6.1. Let $\{\alpha_k, u_k\}_{k=1}^{\infty}$ be as in its statement. Since Q is upper hemicontinuous, then $Q(X)$ is compact. Thus, there exists, for all $i \in N$, $C_i \in \mathbb{R}$ such that $|z_i| \leq C_i$ for all $z \in Q(x)$ and $x \in X$. Define $Z_i = [-C_i, C_i]$ for all $i \in N$ and $Z = \prod_{i \in N} Z_i$.

We start by looking at the probability distribution on $X \times Z$ induced by (α_k, u_k) for all $k \in \mathbb{N}$. Define, for all $k \in \mathbb{N}$, $\pi_k(A \times B) = \alpha_k(A \cap u_k^{-1}(B))$ for all Borel measurable subsets A of X and B of Z. Note that this definition specifies a probability measure π_k on $X \times Z$ since the Borel subsets of $X \times Z$ are generated by sets of the form $A \times B$ as above.

Note that the set $M(X \times Z)$ of Borel probability measures on $X \times Z$ is compact. Hence, we may assume that $\{\pi_k\}_{k=1}^{\infty}$ converges. Let $\pi = \lim_k \pi_k$. The following lemma establishes a property of the support of π.

Lemma 6.2 *The following holds:* $\mathrm{supp}(\pi) \subseteq \mathrm{graph}(Q)$.

Proof. Let $k \in \mathbb{N}$. By Theorem A.25, we have that $\mathrm{supp}(\pi_k) \subseteq \mathrm{cl}(\mathrm{graph}(u_k))$. Since $\mathrm{graph}(u_k) \subseteq \mathrm{graph}(Q)$ and $\mathrm{graph}(Q)$ is closed, then $\mathrm{supp}(\pi_k) \subseteq \mathrm{graph}(Q)$. Hence, it follows by Theorem A.25 again that $\mathrm{supp}(\pi) \subseteq \mathrm{Ls}(\mathrm{supp}(\pi_k)) \subseteq \mathrm{graph}(Q)$. ∎

The following lemma establishes, for every $x \in X$, the existence of a probability measure $\delta(x)$ on Z. This probability measure can be interpreted as the conditional probability measure on payoffs given x. The following notion is used in its statement. Given two separable metric spaces Y and Z, a *transition probability* from Y to Z is a function $\rho : Y \to M(Z)$ such that $y \mapsto \rho(y)(B)$ is measurable for every Borel measurable subset B of Z.

Lemma 6.3   *There exists a transition probability $\delta : X \to M(Z)$ such that:*

1. $\pi(A \times B) = \int_A \delta(x)(B) \mathrm{d}\alpha(x)$ *for all Borel measurable subsets A of X and B of Z, and*
2. $\delta(x)(Q(x)) = 1$ *for all $x \in X$*

   Proof.   Let $\pi_{k,X}$ (resp. π_X) be the marginal of π_k (resp. π) on X for all $k \in \mathbb{N}$. It follows from the definition of π_k that $\pi_{k,X} = \alpha_k$ for all $k \in \mathbb{N}$. Note that $\pi_X = \lim_k \pi_{k,X}$, which can be seen by noting that if $f : X \to \mathbb{R}$ is continuous, then $\hat{f} : X \times Z \to \mathbb{R}$ defined by $\hat{f}(x, z) = f(x)$ for all $(x, z) \in X \times Z$ is also continuous; therefore, $\lim_k \pi_k = \pi$ implies that $\lim_k \int_X f \mathrm{d}\pi_{k,X}(x) = \lim_k \int_{X \times Y} \hat{f}(x, z) \mathrm{d}\pi_k(x, z) = \int_{X \times Y} \hat{f}(x, z) \mathrm{d}\pi(x, z) = \int_X f \mathrm{d}\pi_X(x)$. Hence, $\pi_X = \lim_k \pi_{k,X} = \lim_k \alpha_k = \alpha$.

It then follows by Theorem A.27 that there exists a transition probability $\eta : X \to M(Z)$ such that $\pi(A \times B) = \int_A \eta(x)(B) \mathrm{d}\alpha(x)$ for all Borel measurable subsets A of X and B of Y. Since $\mathrm{supp}(\pi) \subseteq \mathrm{graph}(Q)$, it follows by Theorem A.26 that $1 = \pi(\mathrm{graph}(Q)) = \int_X \eta(x)(Q(x)) \mathrm{d}\alpha(x)$, which in turn implies that $\eta(x)(Q(x)) = 1$ for α-a.e. $x \in X$.

Since Q is well-behaved, it follows by Theorem A.17 that there exists a measurable selection q of Q. Let $T \subseteq X$ be Borel measurable and such that $\alpha(T) = 0$ and $\eta(x)(Q(x)) = 1$ for all $x \in T^c$. Define the transition probability δ by

$$
\delta(x)(B) = \begin{cases} \eta(x)(B) & \text{if } x \in T^c, \\ 1 & \text{if } x \in T \quad \text{and} \quad q(x) \in B, \\ 0 & \text{if } x \in T \quad \text{and} \quad q(x) \notin B \end{cases}
$$

for all $x \in X$ and Borel measurable subset B of Z. Clearly, $\delta(x)(Q(x)) = 1$ for all $x \in X$. Furthermore, because $\alpha(T) = 0$, we have that $\pi(A \times B) = \int_A \delta(x)(B) \mathrm{d}\alpha(x)$ for all Borel measurable subsets A of X and B of Z. ∎

The previous lemma establishes the existence of a probability measure $\delta(x)$ over payoffs for each $x \in X$. By taking its expected value we obtain a payoff vector $\hat{u}(x)$ for each $x \in X$. This will give us a payoff function $\hat{u}$ that has some but not all of the desired properties. For this reason, such function will be modified later on. But, first, we define $\hat{u}$ and analyze its properties.

Let $\hat{u} : X \to Z$ be defined by

$$\hat{u}(x) = \int_Z z\mathrm{d}\delta(x)(z)$$

for all $x \in X$. Note that the compactness of Z and the continuity of $z \mapsto z$ imply that the integral is well-defined.

Lemma 6.4   *The function $\hat{u}$ is measurable and satisfies $\hat{u}(x) \in Q(x)$ for all $x \in X$.*

Proof.   The measurability of $\hat{u}$ follows by Theorem A.26. Since $Q(x)$ is compact and convex for all $x \in X$, the definition of integral and Lemma 6.3 imply that $\hat{u}(x) \in Q(x)$ for all $x \in X$.   ∎

Further properties of $\hat{u}$ are established in the following lemma.

Lemma 6.5   *Let $i \in N$. Then there exists a Borel measurable subset N_i of X_i such that $\alpha_i(N_i) = 0$ and $\hat{u}_i(\alpha) \geq \hat{u}_i(x_i, \alpha_{-i})$ for all $x_i \in N_i^c$. Furthermore, $\lim_k \hat{u}_i(\alpha_k) = \hat{u}_i(\alpha)$.*

Proof.   Let $i \in N$ and B be a Borel measurable subset of X_i. Fix $k \in \mathbb{N}$. Since α_k is a Nash equilibrium of G_k, then $u_i^k(\alpha_k) \geq u_i^k(x_i, \alpha_{-i}^k)$ for all $x_i \in X_i^k$ and $\alpha_i^k(X_i^k) = 1$. Hence, $\alpha_i^k(B)u_i^k(\alpha_k) = \int_B u_i^k(\alpha_k)\mathrm{d}\alpha_i^k(x_i) \geq \int_B u_i^k(x_i, \alpha_{-i}^k)\mathrm{d}\alpha_i^k(x_i)$, i.e.

$$\alpha_i^k(B)\int_X u_i^k \mathrm{d}\alpha_k \geq \int_{B \times X_{-i}} u_i^k \mathrm{d}\alpha_k. \tag{6.1}$$

We next show that

$$\alpha_i(B)\int_X \hat{u}_i \mathrm{d}\alpha \geq \int_{B \times X_{-i}} \hat{u}_i \mathrm{d}\alpha \tag{6.2}$$

by considering several cases.

The first case is where $\alpha_i(\partial B) = 0$, where $\partial B = \mathrm{cl}(B) \setminus \mathrm{int}(B)$ is the boundary of B. Then, $\pi(\partial B \times X_{-i} \times Z) = \alpha_i(\partial B) = 0$. Furthermore, the definition of π_k implies that

$$\int_{B \times X_{-i}} u_i^k \mathrm{d}\alpha_k = \int_{B \times X_{-i} \times Z} z_i \mathrm{d}\pi_k(x, z)$$

for all $k \in \mathbb{N}$. Since $(x, z) \mapsto z_i$ and $(x, z) \mapsto -z_i$ are continuous and bounded, $\pi(\partial B \times X_{-i} \times Z) = \alpha_i(\partial B) = 0$ and $\lim_k \pi_k = \pi$ then Theorem A.19

implies that

$$\lim_k \int_{B \times X_{-i}} u_i^k d\alpha_k = \int_{B \times X_{-i} \times Z} z_i d\pi(x, z). \qquad (6.3)$$

By Lemma 6.3 and Theorem A.26, we have that

$$\int_{B \times X_{-i} \times Z} z_i d\pi(x, z) = \int_{B \times X_{-i}} \left(\int_Z z_i d\delta(x)(z) \right) d\alpha(x) = \int_{B \times X_{-i}} \hat{u}_i d\alpha.$$
$$(6.4)$$

Since $\lim_k \alpha_i^k(B) = \alpha_i(B)$ by Theorem A.19, then (6.2) follows from (6.1), (6.3) and (6.4).

Note that, by taking $B = X_i$ in (6.3) and (6.4), we obtain $\lim_k \hat{u}_i(\alpha_k) = \hat{u}_i(\alpha)$.

The next case we consider is the one where B is closed. Let d denote a metric in X_i and consider, for all $\varepsilon > 0$, $B^\varepsilon = \{x_i \in X_i : d(x_i, B) < \varepsilon\}$. Note that $\partial B^\varepsilon \subseteq \{x_i \in X_i : d(x_i, B) = \varepsilon\}$ and therefore ∂B^ε and $\partial B^{\varepsilon'}$ are disjoint for all $\varepsilon, \varepsilon' > 0$ such that $\varepsilon \neq \varepsilon'$. Thus, the set $\{\varepsilon > 0 : \alpha_i(\partial B^\varepsilon) \geq 1/k\}$ can have at most k elements and, therefore, $\{\varepsilon > 0 : \alpha_i(\partial B^\varepsilon) > 0\} = \cup_{k=1}^\infty \{\varepsilon > 0 : \alpha_i(\partial B^\varepsilon) \geq 1/k\}$ is countable.

Due to the above, there exists a sequence $\{\varepsilon_j\}_{j=1}^\infty$ such that $\lim_j \varepsilon_j = 0$ and $\alpha_i(\partial B^{\varepsilon_j}) = 0$ for all $j \in \mathbb{N}$. Thus, (6.2) holds with B^{ε_j} in place of B for all $j \in \mathbb{N}$. Since B is closed, then $B = \cap_{j=1}^\infty B^{\varepsilon_j}$ and, therefore, $\lim_j \alpha_i(B^{\varepsilon_j}) = \alpha_i(B)$ and $\lim_j \int_{B^{\varepsilon_j} \times X_{-i}} \hat{u}_i d\alpha = \int_{B \times X_{-i}} \hat{u}_i d\alpha$. This implies that (6.2) holds when B is closed.

We finally consider the case of a general B. By Theorem A.18, for all $j \in \mathbb{N}$, there exists a closed set $F_j \subseteq X_i$ such that $\alpha_i(B \setminus F_j) < 1/j$. Furthermore, we may assume that $\{F_j\}_{j=1}^\infty$ is increasing (if not, replace F_j with $F_j' = \cup_{l=1}^j F_l$). Since (6.2) holds with F_j in place of B for all $j \in \mathbb{N}$, $\lim_j \alpha_i(F_j) = \alpha_i(B)$ and $\lim_j \int_{F_j \times X_{-i}} \hat{u}_i d\alpha = \int_{B \times X_{-i}} \hat{u}_i d\alpha$, it follows that implying that (6.2) holds.

To conclude the proof, note that if N_i is the Borel measurable set of $x_i \in X_i$ such that $\hat{u}_i(\alpha) < \hat{u}_i(x_i, \alpha_{-i})$, then $\alpha_i(N_i) > 0$ would imply $\alpha_i(N_i)\hat{u}_i(\alpha) = \int_B \hat{u}_i(\alpha)d\alpha_i(x_i) < \int_{N_i} u_i(x_i, \alpha_{-i})d\alpha_i(x_i)$. But this contradicts (6.2). Thus, $\alpha_i(N_i) = 0$. ∎

The modification of $\hat{u}$ uses the following function. For all $i \in N$, let $w_i : X \to Z_i$ be defined by $w_i(x) = \min_{z \in Q(x)} z_i$ for all $x \in X$.

Lemma 6.6 *For all $i \in N$, w_i well-defined and lower semicontinuous. Furthermore, there exists a measurable $q^i : X \to Z$ such that $q_i^i = w_i$ and $q^i(x) \in Q(x)$ for all $x \in X$.*

Proof. Let $i \in N$. Note that $w_i(x) = -\max_{z \in Q(x)}(-z_i)$ for all $x \in X$. We have that $x \mapsto \max_{z \in Q(x)}(-z_i)$ is well-defined and upper semicontinuous by Theorem A.13 and, therefore, w_i is well-defined and lower semicontinuous.

Define $\psi : X \times Z_i \rightrightarrows Z_{-i}$ by $\psi(x, z_i) = \{z_{-i} \in Z_{-i} : (z_i, z_{-i}) \in Q(x)\}$ for all $(x, z_i) \in X \times Z_i$. Thus, $\operatorname{graph}(\psi) = \operatorname{graph}(Q)$ is closed and ψ is nonempty-valued. Thus, by Theorem A.17, there exists a measurable $g : X \times Z_i \to Z_{-i}$ such that $(z_i, g(x, z_i)) \in Q(x)$ for all $(x, z_i) \in X \times Z_i$. Defining $q^i : X \to Z$ by $q^i(x) = (w_i(x), g(x, w_i(x)))$ for all $x \in X$, it follows that q^i is measurable, $q_i^i = w_i$ and $q^i(x) \in Q(x)$ for all $x \in X$. ∎

For all $i \in N$, let $U^i \subseteq X$ be the (Borel measurable) set of all $x \in X$ such that $x_i \in N_i$ and $x_j \notin N_j$ for all $j \neq i$. Furthermore, let $U = \cup_{i \in N} U^i$. Note that $\alpha(U) = 0$ and $U^i \cap U^j = \emptyset$ for all $i, j \in N$ such that $i \neq j$.

Define $u : X \to Z$ by

$$u(x) = \begin{cases} \hat{u}(x) & \text{if } x \notin U, \\ q^i(x) & \text{if } x \in U^i \end{cases}$$

for all $x \in X$.

We next establish the conclusion of Theorem 6.1.

Proof of Theorem 6.1. It is clear that u is measurable, $u(x) \in Q(x)$ for all $x \in X$ and $\alpha(\{x \in X : u(x) \neq \hat{u}(x)\}) = 0$.

The latter implies that $u(\alpha) = \hat{u}(\alpha)$. Since $\lim_k u_k(\alpha_k) = \hat{u}(\alpha)$ by Lemma 6.5, then $\lim_k u_k(\alpha_k) = u(\alpha)$.

We next show that α is a Nash equilibrium of $G = (X_i, u_i)_{i \in N}$. Let $i \in N$ and $x_i \in X_i$. Letting $U_{x_i} = \{x_{-i} \in X_{-i} : (x_i, x_{-i}) \in U\}$, we have that

$$u_i(x_i, \alpha_{-i}) = \int_{U_{x_i}} u_i(x_i, \cdot) \mathrm{d}\alpha_{-i} + \int_{X_{-i} \setminus U_{x_i}} \hat{u}_i(x_i, \cdot) \mathrm{d}\alpha_{-i}.$$

Since $u(\alpha) = \hat{u}(\alpha)$, it then suffices to show that

$$\hat{u}(\alpha) \geq \int_{U_{x_i}} u_i(x_i, \cdot) \mathrm{d}\alpha_{-i} + \int_{X_{-i} \setminus U_{x_i}} \hat{u}_i(x_i, \cdot) \mathrm{d}\alpha_{-i}. \tag{6.5}$$

We consider two cases. The first case is when $x_i \notin N_i$. In this case, $x_{-i} \in U_{x_i}$ is equivalent to $(x_i, x_{-i}) \in \cup_{j \neq i} U^j$. This implies that $\alpha_{-i}(U_{x_i}) = 0$ and that $\int_{U_{x_i}} u_i(x_i, \cdot) \mathrm{d}\alpha_{-i} + \int_{X_{-i} \setminus U_{x_i}} \hat{u}_i(x_i, \cdot) \mathrm{d}\alpha_{-i} = \int_{X_{-i}} \hat{u}_i(x_i, \cdot) \mathrm{d}\alpha_{-i}$. Since $x_i \notin N_i$, then $\hat{u}(\alpha) \geq \int_{X_{-i}} \hat{u}_i(x_i, \cdot) \mathrm{d}\alpha_{-i}$ by Lemma 6.5 and (6.5) holds.

The second case is when $x_i \in N_i$. In this case, $x_{-i} \in U_{x_i}$ is equivalent to $(x_i, x_{-i}) \in U^i$. Thus, if $x_{-i} \notin U_{x_i}$, then $x_j \in N_j$ for some $j \neq i$. This implies that $\alpha_{-i}(X_{-i} \setminus U_{x_i}) = 0$ and that $\int_{U_{x_i}} u_i(x_i, \cdot) \mathrm{d}\alpha_{-i} + \int_{X_{-i} \setminus U_{x_i}} \hat{u}_i(x_i, \cdot) \mathrm{d}\alpha_{-i} = \int_{U_{x_i}} u_i(x_i, \cdot) \mathrm{d}\alpha_{-i} = \int_{X_{-i}} w_i(x_i, \cdot) \mathrm{d}\alpha_{-i} = w_i(x_i, \alpha_{-i})$.

Since $\mathrm{Li}(X_i^k) = X_i$, let $\{x_i^k\}_{k=1}^\infty$ be such that $\lim_k x_i^k = x_i$ and $x_i^k \in X_i^k$ for all $k \in \mathbb{N}$. Since $w_i \leq u_i^k$ for all $k \in \mathbb{N}$ (by the definition of w_i and by $X_i^k \subseteq X_i$ for all $k \in \mathbb{N}$) and $\mu \mapsto w_i(\mu)$ is lower semicontinuous by Theorem A.23, then

$$w_i(x_i, \alpha_{-i}) \leq \liminf_k w_i(x_i^k, \alpha_{-i}^k) \leq \liminf_k u_i^k(x_i^k, \alpha_{-i}^k).$$

Hence, if $\hat{u}(\alpha) < w_i(x_i, \alpha_{-i})$ then, since $\hat{u}(\alpha) = \lim_k u_k(\alpha_k)$, it follows that $u_i^k(\alpha_k) < u_i^k(x_i^k, \alpha_{-i}^k)$ for all sufficiently large k. But this contradicts the assumption that α_k is a Nash equilibrium of G_k for all $k \in \mathbb{N}$. Thus, $\hat{u}(\alpha) \geq w_i(x_i, \alpha_{-i})$ and (6.5) holds. ∎

The existence of solutions for games with an endogenous sharing rule is easily obtained from Theorem 6.1. In fact, we can approximate players' strategy spaces with finite sets and use the existence result for continuous games (Theorem 2.4) to obtain a sequence of Nash equilibria for the finite-action games. Theorem 6.1 can then be applied by taking a convergent subsequence of such sequence of Nash equilibria.

Theorem 6.7 *Every game with an endogenous sharing rule* $\Gamma = (X_1, \ldots, X_n, Q)$ *has a solution.*

Proof. Let $i \in N$. Since X_i is compact, let $\{x_i^j\}_{j=1}^\infty$ be a dense subset of X_i. Furthermore, for all $i \in N$ and $k \in \mathbb{N}$, let $X_i^k = \{x_i^j\}_{j=1}^k$ and note that $\lim_k \delta_i(X_i^k, X_i) = 0$.

Since Q is well-behaved, it follows by Theorem A.17 that there exists a measurable selection q of Q. Define $u_k = q|_{X_k}$ for all $k \in \mathbb{N}$. Clearly, $u_i^k(x) \in Q(x)$ for all $x \in X_k$ and $k \in \mathbb{N}$.

Fix $k \in \mathbb{N}$. The game $G_k = (X_i^k, u_i^k)_{i \in N}$ is continuous and, hence, has a mixed strategy Nash equilibrium $\alpha_k \in M(X)$ by Theorem 2.4.

Since $M(X)$ is compact, we may assume that the sequence $\{\alpha_k\}_{k=1}^\infty$ converges. It then follows from Theorem 6.1 that Γ has a solution. ∎

6.2 References

Games with an endogenous sharing rule were originally considered in Simon and Zame (1990), where Theorem 6.7 was established. Its proof, as well as Theorem 6.1, are due to Balder (2011).

There are some results on the relationship between the two approaches used to establish the existence of equilibria in discontinuous games, more specifically, between the results in Reny (1999) and those in Simon and Zame (1990). Castro (2011) has shown that every better-reply secure normal-form game is regular (see Theorems 3.17 and 3.19 in Chapter 3) and that every game with an endogenous sharing rule has a measurable selection of the payoff correspondence inducing a regular normal-form game. A stronger conclusion can be obtained in the auction setting of Jackson and Swinkels (2005): For any game with an endogenous sharing rule in the auction setting of Jackson and Swinkels (2005) there exists a measurable selection of the payoff correspondence whose induced normal-form game is better-reply secure. This result was then extended to general games by Carmona and Podczeck (2011).

Games with an endogenous sharing rule are useful in the context of dynamic games. An example of such games is presented in Carmona and Fajardo (2009), where the existence of subgame perfect equilibria in general menu games, known to be sufficient to analyze common agency problems, is established. Menu games are such that a finite set of principals offer menus of contracts to an agent, who chooses after the contracts have been offered. In this setting, discontinuities naturally arise due to the absence, in general, of continuous optimal choices for the agent. However, defining the payoff correspondence at a given strategy profile for the principals as the set of payoffs that arise at all optimal choices of the agent, one obtains a game with an endogenous sharing rule. A similar approach is also useful in the context of the dynamic games considered in Harris, Reny and Robson (1995). In this setting, one may considered the game with an endogenous sharing rule that arises given every possible history by defining the payoff correspondence as the set of payoffs that result from the given history, the strategy played at the current period and any continuation subgame perfect equilibrium. In this context, the convexity of the payoff correspondence is achieved through the presence of public randomization.

Chapter 7

Games With a Continuum of Players

Many economic situations feature a large number of participants. For example, several markets have a large number of buyers and sellers; also, driving in a big city involves the coordination of a large number of drivers. A useful abstraction when formalizing such situations using a normal-form game is to assume that there is a continuum of players. Such an assumption is similar in spirit to standard assumption in economics that prices and quantities can vary continuously and, like those assumptions, is made to make the analysis simpler. This is so partly because, in games with a continuum of players, the impact of each individual on the distribution of actions is negligible, which is also an appropriate property to have when studying interactions between a large number of players.

In this chapter, we consider continuum-of-players games in which players interact in an anonymous way (these games are also called non-atomic games). The latter feature is captured by the assumption that each player is atomless and his payoff depends only on his own action and on the distribution of actions induced by the choices of all players.

We consider a special class of games with a continuum of players, namely, those having a finite set of actions and finitely many possible payoff functions. These two properties are often met in applications and simplify the analysis considerably. A more general analysis can be found in Carmona and Podczeck (2010), on which this chapter is based.

We provide a definition of generalized better-reply security for games with a continuum of players and show that any such game has an equilibrium distribution. Interestingly, this is done by analyzing, using the tools

113

of Chapter 3, a finite-player game defined from the game with a continuum of players.

Furthermore, we establish an equivalence between the equilibrium distributions of the non-atomic game and the Nash equilibria of the corresponding finite-player game. This equivalence implies that, although we do not explicitly analyze the stability of equilibrium distributions for the class of non-atomic games we consider, analogous results to those in Chapter 5 can be obtained for the case of a continuum of players simply by applying their finite-players counterpart.

7.1   Notation and Definitions

We consider games where all players have the same action space X and where each player's payoff depends on her choice and on the distribution of actions induced on X by the choices of all players. The formal setup of the model is as follows.

The action space common to all players is a finite set X. A distributions of actions specifies the fraction of player who choose each of the possible actions $x \in X$. Thus, a distributions of actions is a vector with $|X|$ coordinates, each of which is non-negative and whose sum equals 1. More generally, if Y is a finite set, we let $\Delta(Y) = \{\sigma \in \mathbb{R}_+^{|Y|} : \sum_{y \in Y} \sigma_y = 1\}$ denote the standard $(|Y| - 1)$-dimensional simplex in $\mathbb{R}^{|Y|}$ which, in our context, forms the space of distributions on Y. In particular, $\Delta(X)$ is the space of distribution of actions. We denote a typical element $\pi \in \Delta(Y)$ by $\pi = (\pi_y)_{y \in Y}$.

Each player's payoff depends on his choice and on the distribution of actions and is therefore represented by a real-valued function on $X \times \Delta(X)$. We furthermore assume that players' payoff functions are bounded. Thus, we let the space of players' payoff functions be the space of bounded real-valued functions on $X \times \Delta(X)$, which we endow with the sup-norm and denote by $B(X \times \Delta(X))$.

We only consider non-atomic games with *finite characteristics*, by which we mean that both the action space X and the set of possible players' payoff functions are finite. The latter property amounts to the assumption that players' payoff functions belong to a finite subset U of $B(X \times \Delta(X))$.

So far we have described players' action spaces and the space of their payoff functions. We can complete the description of a game with a continuum of players by specifying, as in the description of a finite-player game, the set of players and a payoff function to each player in that set.

One difference with the finite-player counterpart is the set of players in a non-atomic game is not a finite set but rather an atomless probability space. The main motivation for this formalization is that it implies that each player has no impact on the distribution of actions. Formally, the probability space of players is (T, Σ, ν), where T is the set of players, Σ is a σ-algebra of subsets of T and ν is a probability measure on (T, Σ). Furthermore, the requirement that (T, Σ, ν) is atomless means that for all $A \in \Sigma$ such that $\nu(A) > 0$ there exists $B \in \Sigma$ such that $B \subset A$ and $0 < \nu(B) < \nu(A)$.

A second difference between finite and continuum-of-player games is on how payoff functions are assigned to players. Whereas in the finite-player case, one simply assigns a payoff function to each player, in the presence of a continuum of players, the way that payoff functions are assigned to players must satisfy some measurability property in order for the game to be amendable to analysis. Thus, part of the description of a game with a continuum of players is a measurable function $V : T \to U$. Note that, due to the finiteness of U, the measurability of V means that $V^{-1}(\{u\})$ belongs to Σ for all $u \in U$.

In summary, a *game with a continuum of players* is $G = ((T, \Sigma, \nu), V, X)$ where (T, Σ, ν) is an atomless probability space of players, X is the action space and V is a measurable function from T to U that assigns a payoff function to each player. That is, for all $t \in T$, $V(t)$ is player t's payoff function, with the interpretation that $V(t)(x, \pi)$ is player t's payoff when he plays action x and faces a distribution π in $\Delta(X)$ induced by the actions of all players.

A *strategy of G* is a measurable function $f : T \to X$, i.e. a function f assigning an action to each player such that $f^{-1}(\{x\}) \in \Sigma$ for all $x \in X$. For all $u \in B(X \times \Delta(X))$, let $w_u : \Delta(X) \to \mathbb{R}$ be defined by $w_u(\pi) = \max_{x \in X} u(x, \pi)$ for all $\pi \in \Delta(X)$. A *Nash equilibrium of G* is a strategy f of G such that $V(t)(f(t), \nu \circ f^{-1}) \geq w_{V(t)}(\nu \circ f^{-1})$ for almost all $t \in T$.

We illustrate the above concepts in the following coordination example.

Example 7.1   *Let $T = [0, 1]$ endowed with its Borel σ-algebra and Lebesgue measure. The set of actions consists of two actions $X = \{a, b\}$, which we can think of as being two different locations or brands. All players have the same payoff function which represents a preference for the most popular alternative: $u(x, \pi) = \pi_x$ for all $x \in X$ and $\pi \in \Delta(X)$. Thus, the* *payoff assigning function V is the constant function that assigns u to all players.*

In the above example it is easy to see that the following three strategies are a Nash equilibrium: (1) $f(t) = a$ for all $t \in T$, (2) $f(t) = b$ for all $t \in T$ and (3) $f(t) = a$ for all $t \in [0, 1/2]$ and $f(t) = b$ for all $t \in (1/2, 1]$. Equilibria (1) and (2) achieve coordination since every player chooses the same action. In contrast, equilibrium (3) does not: half of the players choose a and half choose b, which implies that all players are indifferent between the two actions.

The above are only three of the Nash equilibria of the coordination example. We can modify each of these three equilibria to obtain additional Nash equilibria. For example, (4) $f(t) = a$ if $t \in T \cap \mathbb{Q}$ and $f(t) = b$ otherwise, and (5) $f(t) = a$ for all $t \in [0, 1/4] \cup [3/4, 1]$ and $f(t) = b$ for all $t \in (1/4, 3/4)$.

Clearly, the Nash equilibria (4) and (5) can be further modified to produce additional equilibria. Although we can easily identify a continuum of Nash equilibrium, all of the Nash equilibria of Example 7.1 induce one of three distributions τ on $U \times X$: (a) $\tau_{(u,a)} = 1$ (this distribution is induced by Nash equilibrium (1)), (b) $\tau_{(u,b)} = 1$ (this distribution is induced by Nash equilibrium (2)) and (c) $\tau_{(u,a)} = \tau_{(u,b)} = 1/2$ (this distribution is induced by Nash equilibrium (3)).

It follows from above that, while there are many Nash equilibria, there are only three distributions on $U \times X$ that are induced by the payoff-assigning function V and those Nash equilibria. Hence, unlike the case of Nash equilibria, the set of distributions on $U \times X$ induced by the payoff-assigning function and the Nash equilibria is finite in Example 7.1. A similar situation happens regarding the payoff-assigning function: All the above results are correct if we define V by $V(t) = u$ for all $t \in T \cap \mathbb{Q}$ and $V(t) \equiv 0$ otherwise. In other words, the set of Nash equilibria and the set of distributions on $U \times X$ induced by the payoff-assigning function and the Nash equilibria depend only on the fact that the payoff function of almost all players is u (i.e. $\nu(V^{-1}(\{u\})) = 1$).

For the above reasons, which hold generally in the class of non-atomic games we consider, we focus on the distribution that the payoff-assigning function of the game with a continuum of players induces on the space of payoff functions. In particular, a game with a continuum of players is described without explicit reference to the set of players, but rather only by the statistical image of the payoff-assigning function. Thus, for now onwards, a *game with a continuum of players* is simply $\mu \in \Delta(U)$. Moreover, an equilibrium is described only by the statistical image of the payoff-assigning function and the strategy. Formally, a Borel probability measure τ on $B \times X$ is an *equilibrium distribution of* μ if $\tau_B = \mu$ and $\tau(\{(u,x) \in B \times X : u(x, \tau_X) \geq w_u(\tau_X)\}) = 1$.

To represent Example 7.1 in the above way, we set $\mu_u = 1$. Furthermore, the set of equilibrium distributions has three elements τ^1, τ^2 and τ^3 where, as before, $\tau^1_{(u,a)} = 1$, $\tau^2_{(u,b)} = 1$ and $\tau^3_{(u,a)} = \tau^3_{(u,b)} = 1/2$.

A simple but more elaborate example is obtained from Example 7.1 by introducing some players who want to choose the least popular action.

Example 7.2   Let $X = \{a, b\}$, $U = \{u, v\}$, $\mu(u) = 2/3$ and $\mu(v) = 1/3$, where $u(x, \pi) = \pi_x$ and $v(x, \pi) = 1 - \pi_x$ for all $x \in X$ and $\pi \in \Delta(X)$.

It is easy to see that the following are the equilibrium distributions of Example 7.2:

$$\tau_{(u,a)} = 2/3, \ \tau_{(u,b)} = 0, \ \tau_{(v,a)} = 0 \ \text{ and } \ \tau_{(v,b)} = 1/3,$$

$$\tau_{(u,a)} = 0, \ \tau_{(u,b)} = 2/3, \ \tau_{(v,a)} = 1/3 \ \text{ and } \ \tau_{(v,b)} = 0,$$

and, for all $\alpha, \beta \in \mathbb{R}$ such that $0 \leq \alpha \leq 2/3$, $0 \leq \beta \leq 1/3$ and $\alpha + \beta = 1/2$,

$$\tau_{(u,a)} = \alpha, \ \tau_{(u,b)} = 2/3 - \alpha, \ \tau_{(v,a)} = \beta \ \text{ and } \ \tau_{(v,b)} = 1/3 - \beta.$$

Finally note that the payoff functions in Examples 7.1 and 7.2 are continuous. This property implies the existence of at least one equilibrium distribution. Although this is a consequence of Theorem 7.3 below, the main goal of this result is to establish the existence of equilibrium distributions in games with a continuum of players where payoff functions may be discontinuous.

7.2   Existence of Equilibrium Distributions

A difficulty in defining a notion of generalized better-reply security for games with a continuum of players and in applying the techniques developed in Chapter 3 is that the notion of the graph of players' payoff functions is not easily defined. While this is a difficulty for general games with a continuum of players, this problem is easily solved in non-atomic games with finite characteristics. In fact, this class of games allows for a definition of generalized better-reply security that parallels that for games with a finite number of players.

The definition of generalized better-reply security for non-atomic games with finite characteristics is obtained as follows. First, we define the notion of the graph of players' payoff functions. Note that the only continuous variable that affects players' payoffs in a non-atomic game with finite characteristics is the distribution over actions. In particular, the action space is finite and so is the set of possible payoff functions. Hence, we can simply focus on the vectors $(u(x, \pi))_{u \in U, x \in X}$ which lists the payoff for each payoff function and each action as a function of the distribution of actions. Formally, given a game with a continuum of players μ with finite characteristics, let $L = |U \times X|$ and denote a typical element of $\mathbb{R}^L$ by $\alpha = (\alpha(u, x))_{u \in U, x \in X}$. Define

$$\Gamma = \mathrm{cl}(\{(\tau, \alpha) \in \Delta(U \times X) \times \mathbb{R}^L \colon \alpha = (u(x, \tau_X))_{u \in U, x \in X} \text{ and } \tau_U = \mu\}).$$

Second, in the setting of this section, by a well-behaved correspondence from a subset of $\Delta(X)$ into X we mean a upper hemicontinuous correspondence with nonempty and closed values. This is so because the action space, being finite, is not convex (except in the trivial case of a singleton action space).

Third, the generalized payoff secure envelope of players' payoff functions is defined in an analogous way as in the case of finite-player games. Let $N(\pi)$ denote the set of all open neighborhoods of $\pi \in \Delta(X)$. Furthermore, for all $x \in X$, $\pi \in \Delta(X)$ and $O \in N(\pi)$, let $W_O(\pi, x)$ be the set of all well-behaved correspondences $\varphi : O \rightrightarrows X$ that satisfy $(\pi, x) \in \mathrm{graph}(\varphi)$. For all $u \in U$, $x \in X$ and $\pi \in \Delta(X)$, define $\underline{u} \in B(X \times \Delta(X))$ by

$$\underline{u}(x, \pi) = \sup_{O \in N(\pi)} \sup_{\varphi \in W_O(\pi, x)} \inf_{z \in \mathrm{graph}(\varphi)} u(z).$$

We finally turn to the definition of generalized better-reply security for non-atomic games with finite characteristics. Let $\mu \in \Delta(U)$ be a game with a continuum of players with finite characteristics. We say that μ is *generalized better-reply secure* if τ is an equilibrium distribution for every $(\tau, \alpha) \in \Gamma$ such that $\alpha(u, x) \geq w_{\underline{u}}(\tau_X)$ for all $(u, x) \in \mathrm{supp}(\tau)$.

Our main result establishes the existence of equilibrium distributions for non-atomic games with finite characteristics.

Theorem 7.3 *If μ is a generalized better-reply secure game with a continuum of players and with finite characteristics, then μ has an equilibrium distribution.*

The following examples illustrate Theorem 7.3. In Examples 7.1 and 7.2, payoff functions are continuous, which implies that the games in these examples are generalized better-reply secure.

The following example is a non-atomic game with finite characteristics that, while not continuous, is generalized better-reply secure.

Example 7.4   *Let* $X = \{a, b\}$, $U = \{u\}$, $\mu_u = 1$, $0 < \varepsilon < 1/2$ *and* $u(a, \pi) = \pi_a + 2\varepsilon$ *if* $\pi_a \geq 1/2 - \varepsilon$, $u(a, \pi) = \pi_a$ *if* $\pi_a < 1/2 - \varepsilon$ *and* $u(b, \pi) = \pi_b$. *This example modifies Example 7.1 by giving a status-quo effect to action a in the sense that choosing a has a premium of 2ε when at least a fraction of $1/2 - \varepsilon$ is choosing a.*

In Example 7.4, we have that $\underline{u}(b, \pi) = u(b, \pi)$ and

$$\underline{u}(a, \pi) = \begin{cases} \pi_a + 2\varepsilon & \text{if } \pi_a > 1/2 - \varepsilon, \\ \pi_a & \text{if } \pi_a \leq 1/2 - \varepsilon \end{cases}$$

for all $\pi \in \Delta(X)$. We next show that μ is generalized better-reply secure. Let $(\tau, \alpha) \in \Gamma$ be such that $\alpha(u, x) \geq w_{\underline{u}}(\tau_X)$ for all $(u, x) \in \text{supp}(\tau)$. We consider three different cases.

The first case is when $\tau_{X,a} > 1/2 - \varepsilon$. Note that $\alpha(u, b) = \tau_{X,b} = 1 - \tau_{X,a}$. If $(u, b) \in \text{supp}(\tau)$, then $\alpha(u, b) \geq w_{\underline{u}}(\tau_X)$ implies that $1 - \tau_{X,a} \geq \tau_{X,a} + 2\varepsilon$. Therefore, $\tau_{X,a} \leq 1/2 - \varepsilon$ which contradicts $\tau_{X,a} > 1/2 - \varepsilon$. Hence, $(u, b) \notin \text{supp}(\tau)$, which implies that $\tau_{X,b} = 0$ and $\tau_{X,a} = 1$. Therefore, $u(a, \tau_X) = 1 + 2\varepsilon > 0 = u(b, \tau_X)$ and τ is an equilibrium distribution of μ.

The second case is when $\tau_{X,a} < 1/2 - \varepsilon$. Note that, in this case, $\alpha(u, a) = \tau_{X,a}$. If $(u, a) \in \text{supp}(\tau)$, then $\alpha(u, a) \geq w_{\underline{u}}(\tau_X) = 1 - \tau_{X,a}$ implies that $\tau_{X,a} \geq 1 - \tau_{X,a}$ and so $\tau_{X,a} \geq 1/2$, a contradiction. Hence, $\tau_{X,a} = 0$ and $\tau_{X,b} = 1$. Therefore, $u(a, \tau_X) = 0 > 1 = u(b, \tau_X)$ and τ is an equilibrium distribution of μ.

The third and last case is when $\tau_{X,a} = 1/2 - \varepsilon$. In this case, $u(a, \tau_X) = 1/2 + \varepsilon = u(b, \tau_X)$ and τ is an equilibrium distribution.

It follows from the above three cases that μ is generalized better-reply secure and, therefore, μ has an equilibrium distribution.

7.3   Relationship With Finite-Player Games

We will establish Theorem 7.3 using the techniques of Chapter 3. This will be done by defining, for each non-atomic game μ with finite characteristics, a game G with finitely many players with the property that it has approximate equilibria that induce approximate equilibria of μ.

Before proceeding to the actual definition of the finite-player game G, note that if μ is a game with continuum of players and finite characteristics, then $\tau \in \Delta(U \times X)$ is an equilibrium distribution of μ if and only if $\tau_U = \mu$ and, for all $(u, x) \in U \times X$,

$$\tau(u, x) > 0 \quad \text{only if } u(x, \tau_X) \geq w_u(\tau_X). \tag{7.1}$$

Condition (7.1) says that, for each payoff function u, only the actions that are an optimal choice against the distribution of actions τ_X will be played by a strictly positive fraction of the players with payoff function u.

The intuition for the claimed relationship between non-atomic games with finite characteristics and games with finitely many players comes mainly from condition (7.1). In fact, this condition is analogous to the characterization of a mixed strategy equilibrium in a finite-player game which requires that, for each player, only the actions that are an optimal choice against the mixed strategies of the other players will be played with a strictly positive probability. For this analogy to be possible, we need to identify each payoff function in the game with a continuum of players with a player in the game with finitely many players. Furthermore, we need to identify the fraction of players with a given payoff function choosing a given action with the probability of the player in the finite-player game with the given payoff function choosing the given action.

The only difficulty with the above construction is that, in a finite player game, the choice of a player influences τ_X. To avoid this problem, we add a fictitious player, player 0, who will have a strong incentive to choose τ_X. Player 0's role is to guarantee that, in equilibrium, the distribution over actions is the one induced by the choice of players $i = 1, \ldots, n$.

Formally, our construction is as follows. Let μ be a game with a continuum of players and finite characteristics, and let $U = \{u_1, \ldots, u_n\}$ and X be, respectively, the set of payoff functions and actions. Furthermore, let $\mu_i = \mu(\{u_i\})$ for all $i \in \{1, \ldots, n\}$. Define a finite-player game $G = (\Sigma_i, g_i)_{i \in N}$ as follows. The set of players is $N = \{0, \ldots, n\}$ and the action space is common to all players: $\Sigma_i = \Delta(X)$ for all $i \in N$. Player 0's payoff function is defined by

$$g_i(\sigma) = \begin{cases} 1 & \text{if } \sigma_0 = \displaystyle\sum_{i=1}^{n} \mu_i \sigma_i, \\ 0 & \text{otherwise} \end{cases}$$

for all $\sigma \in \Sigma$. For any $i \in \{1, \ldots, n\}$, player i's payoff function is defined by

$$g_i(\sigma) = \sum_{x \in X} \sigma_i(x) u_i(x, \sigma_0)$$

for all $\sigma \in \Sigma$.

The main intuition behind the definition of G is that σ_i can be interpreted as the distribution of actions induced by players with payoff function u_i. Thus, in this interpretation, for all $x \in X$, $\mu_i \sigma_i(x)$ is the fraction of players with payoff function u choosing x. The choice of player 0's payoff function guarantees that in equilibrium (in fact, in approximate equilibrium), player 0 will choose $\sigma_0 = \sum_{i=1}^{n} \mu_i \sigma_i$ which, in the continuum-of-players interpretation, equals the distribution of actions induced by the choices of players with payoff functions in $U = \{u_1, \ldots, u_n\}$. Finally, the choice of player i's payoff function implies that, in equilibrium, the only actions which are played by player i with a strictly positive probability (equivalently, by a strictly positive fraction of players with payoff function u) are those that are a best-reply to σ_0. In the continuum-of-players interpretation, σ_0 is the distribution of actions induced by the choices of players with payoff functions in U and, therefore, the latter property implies that the distribution $(\mu_i \sigma_i(x))_{i \in \{1, \ldots, n\}, x \in X}$ is an equilibrium distribution of μ.

The above discussion can be expressed formally as follows. Given $\sigma \in \Sigma$, let $\tau_\sigma \in \Delta(U \times X)$ be defined by

$$\tau_\sigma(u_i, x) = \mu_i \sigma_i(x)$$

for all $i \in \{1, \ldots, n\}$ and $x \in X$. Note that $\tau_{\sigma, U} = \mu$ since $\tau_{\sigma, U}(u_i) = \sum_{x \in X} \mu_i \sigma_i(x) = \mu_i$ for all $i \in \{1, \ldots, n\}$. Furthermore, $\tau_{\sigma, X} = \sum_{i=1}^{n} \mu_i \sigma_i$.

The next two results show that the set of Nash equilibria of G is equivalent to the set of equilibrium distributions of μ. These results, in particular, imply that limit results for non-atomic games with finite characteristics can be established using limit results for finite-player games.

Theorem 7.5 *If σ is a Nash equilibrium of G then τ_σ is an equilibrium distribution of μ.*

Proof. Let σ be a Nash equilibrium of G. Then we must have $\sigma_0 = \sum_{i=1}^{n} \mu_i \sigma_i$; hence, $\sigma_0 = \tau_{\sigma, X}$. This implies that, for any $i \in \{1, \ldots, n\}$, $g_i(\sigma) = \sum_{x \in X} \sigma_i(x) u_i(x, \tau_{\sigma, X})$; hence, if $\tau_\sigma(u_i, x) > 0$ for some $x \in X$, then $\sigma_i(x) > 0$ and, since σ is a Nash equilibrium, we have $u_i(x, \tau_{\sigma, X}) \geq w_{u_i}(\tau_{\sigma, X})$. Since this conclusion holds for all $i \in \{1, \ldots, n\}$ and $x \in X$, it follows that τ_σ satisfies (7.1). Since $\tau_{\sigma,U} = \mu$, it follows that τ_σ is an equilibrium distribution of μ. ∎

We next show that a converse of Theorem 7.5 also holds. To this end, given $\tau \in \Delta(U \times X)$, let $\sigma_\tau \in \Sigma$ be defined by

$$\sigma_{\tau,i}(x) = \begin{cases} \tau_X & \text{if } i = 0, \\ \tau(u_i, x)/\mu_i & \text{otherwise} \end{cases}$$

for all $i \in N$ and $x \in X$.

Theorem 7.6 *If τ is an equilibrium distribution of μ then σ_τ is a Nash equilibrium of G.*

Proof. Let τ be an equilibrium distribution of μ. Note that, for all $x \in X$, $\sum_{i=1}^n \mu_i \sigma_{\tau,i}(x) = \sum_{i=1}^n \tau(u_i, x) = \tau_X(x) = \sigma_{\tau,0}(x)$. Hence, $\sum_{i=1}^n \mu_i \sigma_{\tau,i} = \sigma_{\tau,0}$ and, therefore, $g_0(\sigma_\tau) = \max_{\sigma_0' \in \Sigma_0} g_0(\sigma_0', \sigma_{\tau,-i})$.

Fix $i \in \{1, \ldots, n\}$. Since $\sigma_{\tau,0} = \tau_X$, we have that $g_i(\sigma_\tau) = \sum_{x \in X} \sigma_{\tau,i}(x) u_i(x, \tau_X)$. If $\sigma_{\tau,i}(x) > 0$ for some $x \in X$, then $\tau(u_i, x) > 0$ and, since τ is an equilibrium distribution, we have that $u_i(x, \tau_X) \geq w_{u_i}(\tau_X)$ by (7.1). Thus, $g_i(\sigma_\tau) = \sum_{x:\sigma_{\tau,i}(x)>0} \sigma_{\tau,i}(x) u_i(x, \tau_X) \geq w_{u_i}(\tau_X) = \max_{\sigma_i' \in \Sigma_i} g_i(\sigma_i', \sigma_{\tau,-i})$. ∎

7.4 Proof of the Existence Theorem for Non-atomic Games

Despite the above relationship between the equilibrium sets of G and μ, in general, we cannot establish the existence of an equilibrium distribution of μ using, first, Theorem 3.2 to obtain a Nash equilibrium of G and, second, Theorem 7.5 to obtain an equilibrium distribution of μ. The reason is that the generalized better-reply security of μ does not imply that of G. To show that G is generalized better-reply secure, we need to consider $(\sigma^*, g^*) \in \text{cl}(\text{graph}(g))$ such that $g_i^* \geq w_{g_i}(\sigma^*)$ for all $i \in \{0, \ldots, n\}$ and show that σ^* is a Nash equilibrium of G. We can then consider a sequence $\{\sigma_k\}_{k=1}^\infty$ such that $\lim_k(\sigma_k, g(\sigma_k)) = (\sigma^*, g^*)$ and define, for all $k \in \mathbb{N}$, τ_{σ_k} and $\alpha_k = (u(x, \tau_{\sigma_k,X}))_{u \in U, x \in X}$. From the latter sequence, we can obtain (τ^*, α^*) such that, taking subsequence if needed, $\lim_k(\tau_{\sigma_k}, \alpha_k) = (\tau^*, \alpha^*)$. While it is possible to show that $\sigma_{\tau^*} = \sigma^*$, $(\tau^*, \alpha^*) \in \Gamma$ and that $\sum_{x \in X} \tau^*(u_i, x)\alpha^*(u, x) \geq w_{u_i}(\tau_X^*)$ for all $i \in \{1, \ldots, n\}$, in general it is not true that $\alpha^*(u, x) \geq w_{u_i}(\tau_X^*)$ for all $(u_i, x) \in \text{supp}(\tau^*)$. The failure of the latter property prevent us to conclude, using the generalized better-reply security of μ, that τ^* is an equilibrium distribution of μ and that $\sigma^* = \sigma_{\tau^*}$ is a Nash equilibrium of G.

Nevertheless, the game G is still useful to establish the existence of an equilibrium distribution of μ, and this will be done by extending the arguments used in Chapter 3.

The first step of the argument consists in showing that, for all continuous $f \in F(\underline{G})$ such that f_i depends only on σ_{-i} and satisfies $f_i < w_{\underline{g}_i}$ for all $i \in \{0, \dots, n\}$, the finite-player G has a f-equilibrium σ in which each player $j = 1, \dots, n$ assigns strictly positive probability only to actions that provide at least a payoff of $f_j(\sigma_{-j})$.

Lemma 7.7 *Let $\underline{G} = (\Sigma_i, \underline{g}_i)_{i=0}^n$ and $f \in F(\underline{G})$ be continuous and such that $f_i(\sigma) = f_i(\sigma_i', \sigma_{-i})$ and $f_i(\sigma_{-i}) < w_{\underline{g}_i}(\sigma_{-i})$ for all $i \in \{0, \dots, 1\}$, $\sigma_i' \in \Sigma_i$ and $\sigma \in \Sigma$. Then, G has an f-equilibrium σ such that $g_i(x, \sigma_{-i}) > f_i(\sigma_{-i})$ for all $i \in \{1, \dots, n\}$ and $x \in \mathrm{supp}(\sigma_i)$.*

Proof. For all $i \in \{1, \dots, n\}$, let $A_i : \Sigma \rightrightarrows X$ be defined by

$$A_i(\sigma) = \{x \in X : g_i(x, \sigma_{-i}) > f_i(\sigma_{-i})\}$$

for all $\sigma \in \Sigma$. Furthermore, let $A_0 : \Sigma \rightrightarrows \Sigma_0$ be defined by

$$A_0(\sigma) = \{\sigma_0 \in \Sigma_0 : g_0(\sigma_0, \sigma_{-0}) > f_i(\sigma_{-i})\}$$

for all $\sigma \in \Sigma$. Then, define $B : \Sigma \rightrightarrows \Sigma$ by setting, for all $\sigma \in \Sigma$,

$$B(\sigma) = A_0(\sigma) \times \mathrm{co}\left(\prod_{i=1}^n A_i(\sigma)\right).$$

Using analogous arguments as in the proof of Lemma 3.5, we can show that B has a fixed point σ^*. Thus, for all $i \in \{1, \dots, n\}$, $\sigma_i^*(x) > 0$ implies that $g_i(x, \sigma_{-i}^*) > f_i(\sigma^*)$ as desired. Finally, this property also implies that $u_i(\sigma^*) > f_i(\sigma^*)$ for all $i \in \{1, \dots, n\}$. Since $\sigma_0^* \in A_0(\sigma^*)$, then we also have $u_0(\sigma^*) > f_0(\sigma^*)$ and, therefore, σ^* is a f-equilibrium of G. ∎

Lemma 7.7 implies that μ also have f-equilibria, as stated in Lemma 7.9. The latter result requires a relationship of $w_{\underline{g}_i}$ and $w_{\underline{u}_i}$ for all $i \in \{1, \dots, n\}$ established in the following lemma.

Lemma 7.8 *The following holds:*

1. $w_{\underline{g}_0}(\sigma_{-0}) = 1$ *for all* $\sigma_{-0} \in \Sigma_{-0}$ *and*
2. $w_{\underline{g}_i}(\sigma_{-i}) \geq w_{\underline{u}_i}(\sigma_0)$ *for all* $i \in \{1, \dots, n\}$ *and* $\sigma_{-i} \in \Sigma_{-i}$.

Proof. We first show that $w_{g_0}(\sigma_{-0}) = 1$ for all $\sigma_{-0} \in \Sigma_{-0}$. In fact, for any $\sigma \in \Sigma$, letting $V_{\sigma_{-0}} = \Sigma_{-0}$ and $\varphi_0 : V_{\sigma_{-0}} \to \Sigma_0$ be defined by $\varphi_0(\sigma_{-0}) = \sum_{i=1}^{n} \mu_i \sigma_i$ for all $\sigma_{-0} \in V_{\sigma_{-0}}$, we have that $V_{\sigma_{-0}}$ is an open neighborhood of σ_{-0}, φ_0 is a continuous function and $\inf_{z \in \mathrm{graph}(\varphi)} g_0(z) = 1$. Since $g_0(z) \leq 1$ for all $z \in \Sigma$, it follows that $\underline{g}_0(\sigma) = 1$ and, hence, $w_{g_0}(\sigma_{-0}) = 1$ for all $\sigma_{-0} \in \Sigma_{-0}$.

We next claim that $w_{g_i}(\sigma_{-i}) \geq w_{\underline{u}_i}(\sigma_0)$ for all $i \in \{1, \ldots, n\}$ and $\sigma_{-i} \in \Sigma_{-i}$. Let $i \in \{1, \ldots, n\}$, $\sigma_{-i} \in \Sigma_{-i}$ and $\varepsilon > 0$ be given. Let $\bar{x} \in X$, $V_{\sigma_0} \in N(\sigma_0)$ and $\varphi_i : V_{\sigma_0} \rightrightarrows X$ be a well-behaved correspondence such that $\bar{x} \in \varphi_i(\sigma_0)$ and

$$\inf_{z \in \mathrm{graph}(\varphi_i)} u_i(z) > w_{\underline{u}_i}(\sigma_0) - \varepsilon.$$

Define $V_{\sigma_{-i}} = V_{\sigma_0} \times \Sigma_{-i}$ and $\psi_i : V_{\sigma_{-i}} \rightrightarrows \Sigma_i$ by

$$\psi_i(\sigma'_{-i}) = \mathrm{co}(\varphi_i(\sigma'_0))$$

for all $\sigma'_{-i} \in V_{\sigma_{-i}}$. Then $V_{\sigma_{-i}} \in N(\sigma_{-i})$ and $\psi_i \in W_{V_{\sigma_{-i}}}(1_{\bar{x}}, \sigma_{-i})$. Take $\sigma' \in \mathrm{graph}(\psi_i)$. Then, $\sigma'_0 \in V_{\sigma_0}$ and $\sigma'_i = \sum_{x \in \varphi_i(\sigma'_0)} \sigma'_i(x) 1_x$ by Theorem A.6. Hence,

$$g_i(\sigma') = \sum_{x \in \varphi_i(\sigma'_0)} \sigma'_i(x) u_i(x, \sigma'_0) \geq \sum_{x \in \varphi_i(\sigma'_0)} \sigma'_i(x) \inf_{z \in \mathrm{graph}(\varphi_i)} u_i(z)$$

$$= \inf_{z \in \mathrm{graph}(\varphi_i)} u_i(z) > w_{\underline{u}_i}(\sigma_0) - \varepsilon.$$

It then follows that

$$w_{g_i}(\sigma_{-i}) \geq \underline{g}_i(1_{\bar{x}}, \sigma_{-i}) \geq \inf_{\sigma' \in \mathrm{graph}(\psi_i)} g_i(\sigma') \geq w_{\underline{u}_i}(\sigma_0) - \varepsilon.$$

Since $\varepsilon > 0$ is arbitrary, it follows that $w_{g_i}(\sigma_{-i}) \geq w_{\underline{u}_i}(\sigma_0)$. ∎

Lemma 7.9 establishes the existence of approximate equilibria of μ. It requires the following definitions. Let $\tilde{F}(\mu)$ be the set of all bounded functions $f = (f_1, \ldots, f_n) : \Delta(X) \to \mathbb{R}^n$ that satisfy and $f_i(\pi) < w_{u_i}(\pi)$ for all $\pi \in \Delta(X)$ and $i \in \{1, \ldots, n\}$. We say that $\tau \in \Delta(U \times X)$ is a *f-equilibrium distribution of* μ if $\tau_U = \mu$ and, for all $(i, x) \in \{1, \ldots, n\} \times X$, $\tau(u_i, x) > 0$ implies $u_i(x, \tau_X) \geq f_i(\tau_X)$.

Lemma 7.9   *If $f \in \tilde{F}(\mu)$ is continuous, then μ has an f-equilibrium.*

Proof. Let $f \in \tilde{F}(\mu)$ be continuous. For all $i \in \{1, \ldots, n\}$, define $\hat{f}_i : \Sigma_{-i} \to \mathbb{R}$ by $\hat{f}_i(\sigma_{-i}) = f_i(\sigma_0)$ for all $\sigma_{-i} \in \Sigma_{-i}$. Let $0 < \varepsilon < 1$ and define $\hat{f}_0 : \Sigma_{-0} \to \mathbb{R}$ by $\hat{f}_0(\sigma_{-0}) = \varepsilon$ for all $\sigma_{-0} \in \Sigma_{-0}$. Thus $\hat{f}$ is continuous and, by Lemma 7.8, $\hat{f}_i < w_{\underline{g}_i}$ for all $i \in \{0, \ldots, n\}$. Hence, by Lemma 7.7, there exists an $\hat{f}$-equilibrium $\sigma \in \Sigma$ of G such that $g_i(x, \sigma_{-i}) > \hat{f}_i(\sigma_{-i})$ for all $i \in \{1, \ldots, n\}$ and $x \in \operatorname{supp}(\sigma_i)$.

Consider τ_σ. Since $u_0(\sigma) > \hat{f}_0(\sigma_{-0}) > 0$, then $u_0(\sigma) = 1$. Hence, $\sigma_0 = \sum_{i=1}^{n} \mu_i \sigma_i = \tau_{\sigma,X}$. It then follows that if $(i, x) \in \{1, \ldots, n\} \times X$ is such that $\tau_\sigma(u_i, x) > 0$, then $x \in \operatorname{supp}(\sigma_i)$ and so $u_i(x, \tau_{\sigma,X}) = g_i(x, \sigma_0) = g_i(x, \sigma_{-i}) > \hat{f}_i(\sigma_{-i}) = f_i(\sigma_0) = f_i(\tau_{\sigma,X})$. Since this property holds for all $(i, x) \in \{1, \ldots, n\} \times X$ such that $\tau_\sigma(u_i, x) > 0$ and we have that $\tau_{\sigma,U} = \mu$, then τ_σ is an f-equilibrium distribution of μ. ∎

Lemma 7.9 implies that, to establish the existence of an equilibrium distribution of μ, it suffices to show that limit points of f_k-equilibria of μ, with f_k suitably approaching $(w_{\underline{u}_i})_{i=1}^{n}$, are equilibrium distributions of μ. Thus, we simply need to adapt the arguments of Lemma 3.6 to complete the proof of Theorem 7.3.

Proof of Theorem 7.3. Let, for all $i \in \{1, \ldots, n\}$, $\{v_i^k\}_{k=1}^{\infty}$ be a sequence of continuous real-valued functions on $\Delta(X)$ such that $v_i^k(\pi) \leq w_{\underline{u}_i}(\pi)$ and $\liminf_k v_i^k(\pi_k) \geq w_{\underline{u}_i}(\pi)$ for all $k \in \mathbb{N}$, $\pi \in \Delta(X)$ and all sequences $\{\pi_k\}_{k=1}^{\infty}$ converging to π (the existence of this sequence follows from Lemma 3.4 and Theorem A.1). Moreover, for all $k \in \mathbb{N}$ and $i \in \{1, \ldots, n\}$, let $f_i^k : \Delta(X) \to \mathbb{R}$ be defined by $f_i^k(\pi) = v_i^k(\pi) - 1/k$ for all $\pi \in \Delta(X)$ and note that $f_k = (f_1^k, \ldots, f_n^k) \in \tilde{F}(\mu)$.

Lemma 7.9 implies that, for all $k \in \mathbb{N}$, μ has a f_k-equilibrium $\tau_k \in \Delta(U \times X)$. Let $\alpha_k = (u(x, \tau_{k,X}))_{u \in U, x \in X}$ for all $k \in \mathbb{N}$. Since $\Delta(U \times X)$ is compact and each $u \in U$ is bounded, we may assume that $\{(\tau_k, \alpha_k)\}_{k=1}^{\infty}$ converges. Let $(\tau^*, \alpha^*) = \lim_k (\tau_k, \alpha_k)$.

Let $(u, x) \in \operatorname{supp}(\tau^*)$. Since $\tau^*(u, x) = \lim_k \tau_k(u, x)$, it follows that $(u, x) \in \operatorname{supp}(\tau_k)$ for all k sufficiently large. Since τ_k is a f_k-equilibrium distribution of μ,

$$\alpha^*(u, x) = \lim_k u_i(x, \tau_{k,X}) \geq \liminf_k f_i(\tau_{k,X}) \geq w_{\underline{u}_i}(\tau_X^*).$$

Thus, it follows that $\alpha^*(u, x) \geq w_{\underline{u}_i}(\tau_X^*)$ for all $(u, x) \in \operatorname{supp}(\tau^*)$. Since μ is generalized better-reply secure and we clearly have that $(\tau^*, \alpha^*) \in \Gamma$, it follows that τ^* is an equilibrium distribution of μ. ∎

7.5   References

The notion of generalized better-reply security for games with a continuum of players, as well as Theorem 7.3, is due to Carmona and Podczeck (2010).

We decided to present the case of games with a continuum of players with finite characteristics for simplicity. Papers that address the existence of equilibrium in general games with a continuum of players and/or discuss the relationship between these games with finite-player games include Al-Najjar (2008), Balder (2002), Carmona and Podczeck (2009), Carmona and Podczeck (2010), Carmona and Podczeck (2012), Khan and Sun (1999), Khan, Rath, and Sun (1997), Mas-Colell (1984), Rashid (1983), Rath (1992) and Schmeidler (1973). Many of these papers and other related work is surveyed in Khan and Sun (2002).

Appendix A

Mathematical Appendix

We collect some mathematical results in this appendix. These results have been specialized to the case of metric spaces and, sometimes, to the case of compact metric spaces. For this reason, some of them hold under more general assumptions that those assumed here.

The first result is an approximation result for lower semicontinuous functions in terms of continuous functions (see Reny, 1999, Lemma 3.5).

Theorem A.1 *Let X be a compact metric space and $g : X \to \mathbb{R}$ be a lower semicontinuous function. Then there exists a sequence $\{g_k\}_{k=1}^{\infty}$ of continuous real-valued functions on X such that $g_k \leq g$ for all $k \in \mathbb{N}$ and $\liminf_k g_k(x_k) \geq g(x)$ for all $x \in X$ and all sequences $\{x_k\}_{k=1}^{\infty} \subseteq X$ converging to x.*

Let (X, d) be a compact metric space and F and F' be closed subsets of X. Recall that the Hausdorff distance between F and F' is given by

$$\delta(F, F') = \max\{\sup_{x \in F} d(x, F'), \sup_{x' \in F'} d(x', F)\},$$

where $d(z, A) = \inf\{d(z, y) : y \in A\}$ for all $z \in X$ and $A \subseteq X$. Furthermore, given a sequence $\{E_k\}_{k=1}^{\infty} \subseteq X$, the *topological limsup* of $\{E_k\}_{k=1}^{\infty}$, denoted $\mathrm{Ls}(E_k)$, consists of the points $z \in Z$ such that, for every neighborhood V of z, there exist infinitely many $k \in \mathbb{N}$ with $V \cap E_k \neq \emptyset$. Moreover, the *topological liminf* of $\{E_k\}_{k=1}^{\infty}$, denoted $\mathrm{Li}(E_k)$, consists of the points $z \in Z$ such that, for every neighborhood V of z, we have $V \cap E_k \neq \emptyset$ for all but finitely many $k \in \mathbb{N}$.

The following result presents a characterization of convergence according to the Hausdorff distance (see Aliprantis and Border, 2006, Theorem 3.93, p. 121).

Theorem A.2   *If X is a compact metric space, F is a closed subset of X and $\{F_k\}_{k=1}^{\infty}$ is a sequence of closed subsets of X, then $\lim_k \delta(F_k, F) = 0$ if and only if $F = \mathrm{Ls}(F_k) = \mathrm{Li}(F_k)$.*

Let X be a compact metric space and $\{U_k\}_{k=1}^{m}$ be an open cover of X, i.e. U_k is open for all $k = 1, \ldots, m$ and $X \subseteq \cup_{k=1}^{m} U_k$. A collection $\{f_k\}_{k=1}^{m}$ of continuous functions from X into $[0,1]$ is a *partition of unity subordinated to* $\{U_k\}_{k=1}^{m}$ if $\sum_{k=1}^{m} f_k(x) = 1$ for all $x \in X$ and $f_k(x) = 0$ for all $k = 1, \ldots, m$ and $x \notin U_k$. The following result establishes the existence of a partition of unity in compact metric spaces (see Aliprantis and Border, 2006, Lemma 2.92, p. 67).

Theorem A.3   *Let X be a compact metric space and $\{U_k\}_{k=1}^{m}$ be an open cover of X. Then there exists a partition of unity subordinated to $\{U_k\}_{k=1}^{m}$.*

Let X and Y be metric spaces and $\Psi : X \rightrightarrows Y$ be a correspondence. We say that Ψ is *upper hemicontinuous* if, for all $x \in X$ and all open $U \subseteq Y$ such that $\Psi(x) \subseteq U$, there exists a neighborhood V of x such that $\Psi(x') \subseteq U$ for all $x' \in V$. Furthermore, Ψ has *nonempty (resp. convex, closed, compact) values* if $\Psi(y)$ is nonempty (resp. convex, closed, compact) for all $y \in Y$.

The following result characterizes upper hemicontinuity in terms of sequences (see Aliprantis and Border, 2006, Corollary 17.17, p. 564).

Theorem A.4   *Let X and Y be metric spaces and $\Psi : X \rightrightarrows Y$ be a compact-valued correspondence. Then Ψ is upper hemicontinuous if and only if for every sequence $\{(x_k, y_k)\}_{k=1}^{\infty}$ such that $\lim_k x_k = x$ for some $x \in X$, $x_k \in X$ and $y_k \in \Psi(x_k)$ for all $k \in \mathbb{N}$, the sequence $\{y_k\}_{k=1}^{\infty}$ has a limit point in $\Psi(x)$.*

A correspondence $\Phi : X \rightrightarrows Y$ is *closed* if graph(Φ) is a closed subset of $X \times Y$. A correspondences into a compact metric space is closed if and only if it is upper hemicontinuous and closed-valued (see Aliprantis and Border, 2006, Theorem 17.11, p. 561).

Theorem A.5   *Let X be a metric space and Y be a compact metric space. Then $\Phi : X \rightrightarrows Y$ is closed if and only if Φ is upper hemicontinuous and closed-valued.*

Given a correspondence $\Psi : X \rightrightarrows Y$, let $\mathrm{co}\Psi : X \rightrightarrows Y$ defined by $\mathrm{co}\Psi(x) = \mathrm{co}(\Psi(x))$ for all $x \in X$. The following result provides sufficient conditions for coΨ to be upper hemicontinuous and compact-valued (see Lemma 5.29 and Theorem 17.35 in Aliprantis and Border (2006)).

Theorem A.6 *If X is a metric space, Y is a compact subset of a locally convex metric vector space space and $\Psi : X \rightrightarrows Y$ is upper hemicontinuous and compact-valued, then coΨ is also upper hemicontinuous and compact-valued.*

Let X and Y be metric spaces and $\Psi : X \rightrightarrows Y$ be a correspondence. We say that Ψ is *lower hemicontinuous at* $x \in X$ if, for all open $U \subseteq Y$ such that $\Psi(x) \cap U \neq \emptyset$, there exists a neighborhood V of x such that $\Psi(x') \cap U \neq \emptyset$ for all $x' \in V$. If Ψ is lower hemicontinuous at x for all $x \in X$, then we say that Ψ is *lower hemicontinuous*.

The following result provides a characterization of lower hemicontinuity (see Aliprantis and Border, 2006, Lemma 17.5, p. 559). Furthermore, it asserts that the closure of a lower hemicontinuous correspondence is itself lower hemicontinuous (see Aliprantis and Border, 2006, Lemma 17.22). By the closure of a correspondence $\Psi : X \rightrightarrows Y$, we mean the correspondence cl$\Psi : X \rightrightarrows Y$ defined by cl$\Psi(x) = \text{cl}(\Psi(x))$ for all $x \in X$.

Theorem A.7 *Let X and Y be metric spaces and $\Psi : X \rightrightarrows Y$ be a correspondence.*

1. *Ψ is lower hemicontinuous if and only if $\{x \in X : \Psi(x) \subseteq F\}$ is closed for each closed subset F of Y.*
2. *If Ψ is lower hemicontinuous then clΨ is also lower hemicontinuous.*

The following result considers convex combinations of correspondences (see Barelli and Soza, 2010, Lemma 5.2).

Theorem A.8 *Let X be a metric space, Y be a metric vector space, $m \in \mathbb{N}$ and, for all $1 \leq j \leq m$, $\Psi_j : X \rightrightarrows Y$ be an upper hemicontinuous compact-valued correspondence and $g_j : X \to [0,1]$ be continuous and such that $\sum_{j=1}^m g_j(x) = 1$ for all $x \in X$.*
Then the correspondence $x \mapsto \sum_{j=1}^m g_j(x)\Psi_j(x)$ is upper hemicontinous with compact-valued. Furthermore, if Ψ_j is convex-valued for all $1 \leq j \leq m$, then $x \mapsto \sum_{j=1}^m g_j(x)\Psi_j(x)$ is also convex-valued.

Let X be a metric space. A subset $T \subseteq X$ is *nowhere dense* if $\text{int}(\text{cl}(T)) = \emptyset$, it is *first category in X* if T is a countable union of nowhere dense sets and is *second category in X* if T is not first category.

The following result states that the set of point where a compact-valued, upper hemicontinuous correspondence fails to be lower hemicontinuous is a first category subset of its domain (see Fort, 1951, Theorem 2).

Theorem A.9   *Let X and Y be metric spaces and $\Psi : X \rightrightarrows Y$ be a upper hemicontinuous correspondence with nonempty compact values. Then there exists a first category set $T \subseteq X$ such that Ψ is lower hemicontinuous at every point in T^c.*

We next recall Baire's category theorem which provides a sufficient condition for a metric space to be second category in itself (see Aliprantis and Border, 2006, Theorem 3.47, p. 94).

Theorem A.10   *Every complete metric space is second category in itself.*

Weierstrass Theorem establishes the existence of a maximum for continuous real-valued functions with compact domain (see Aliprantis and Border, 2006, Corollary 2.35, p. 40).

Theorem A.11   *If X is a compact metric space and $f : X \to \mathbb{R}$ is continuous, then there exists $x^* \in X$ such that $f(x^*) \geq f(x)$ for all $x \in X$.*

The following two results also consider optimization problems. The first result shows that if the objective function is jointly lower semicontinuous on the decision variable and on a parameter, then the value function is a lower semicontinuous function of the parameter (see Aliprantis and Border, 2006, Lemma 17.29, p. 569).

Theorem A.12   *If X is a compact metric space, Y is a metric space and $f : X \times Y \to \mathbb{R}$ is lower semicontinuous, then $y \mapsto \sup_{x \in X} f(x,y)$ is lower semicontinuous.*

The next result considers the case where the objective function is upper semicontinuous and allows for the presence of a constraint set that varies in an upper hemicontinuous way with the parameter. In this case, the value function is upper semicontinuous (see Aliprantis and Border, 2006, Lemma 17.30, p. 569).

Theorem A.13   *If X is a compact metric space, Y is a metric space, $\varphi : Y \rightrightarrows X$ is closed and $f : X \times Y \to \mathbb{R}$ is upper semicontinuous, then $\sup_{x \in \varphi(y)} f(x,y) = \max_{x \in \varphi(y)} f(x,y)$ and $y \mapsto \max_{x \in \varphi(y)} f(x,y)$ is upper semicontinuous.*

Given a metric space X and a correspondence $\Phi : X \rightrightarrows X$, $x \in X$ is a *fixed point of* Φ if $x \in \Phi(x)$. The following is Cauty's generalization of Schauder's fixed point theorem (see Aliprantis and Border, 2006, Corollary 17.56, p. 583 and Cauty, 2001).

Theorem A.14 *If X is a nonempty, compact and convex subset of a metric vector space and $\Phi : X \rightrightarrows X$ is upper hemicontinuous with nonempty, convex and compact values, then Φ has a fixed point.*

The above result relies, in particular, on the assumption that the correspondence Φ is closed. The following result, Browder's fixed point theorem, considers instead the case where Φ has open lower sections (see Browder, 1968, Theorem 1).

Theorem A.15 *Let X be a nonempty, compact and convex subset of a metric vector space. If $\Phi : X \rightrightarrows X$ is such that $\Phi^{-1}(y) = \{x \in X : y \in \Phi(x)\}$ is open for all $y \in X$ and $\Phi(x)$ is nonempty and convex for all $x \in X$, then Φ has a fixed point.*

Let X and Y be metric spaces and $\Phi : X \rightrightarrows Y$ be a correspondence. A function $f : X \to Y$ is a *continuous selection of* Φ if f is continuous and $f(x) \in \Phi(x)$ for all $x \in X$. The following result, Michael's selection theorem, establishes the existence of continuous selections for lower hemicontinuous correspondences with nonempty, closed and convex values (see Aliprantis and Border, 2006, Theorem 17.66, p. 589).

Theorem A.16 *Let X be a compact metric space, $m \in \mathbb{N}$ and $\Phi : X \rightrightarrows \mathbb{R}^m$ be a lower hemicontinuous correspondence with nonempty, closed and convex values. Then Φ has a continuous selection.*

Let X be a metric space. A nonempty family $\mathcal{A}$ of subsets of X is a *σ-algebra* if: (a) $A \in \mathcal{A}$ implies that $A^c \in \mathcal{A}$ and (b) $\{A_n\}_{n=1}^{\infty} \subseteq \mathcal{A}$ implies $\cup_{n=1}^{\infty} A_n \in \mathcal{A}$. Given a nonempty family $\mathcal{C}$ of subsets of X, the *σ-algebra generated by $\mathcal{C}$* is the smallest σ-algebra containing $\mathcal{C}$ and is denoted by $\sigma(\mathcal{C})$. When $\mathcal{C}$ is the family of open subsets of X, $\sigma(\mathcal{C})$ is denoted $\mathcal{B}(X)$ and called the Borel σ-algebra of X.

We next consider a measurable selection result. Let X and Y be metric spaces and $\Phi : X \rightrightarrows Y$ be a correspondence. A function $f : X \to Y$ is a *measurable selection of* Φ if f is measurable (i.e. $f^{-1}(A) \in \mathcal{B}(X)$ for all $A \in \mathcal{B}(Y)$) and $f(x) \in \Phi(x)$ for all $x \in X$. Closed correspondences with nonempty values have a measurable selection (see Theorem 18.20, Lemma 18.2 and Theorem 18.13 in Aliprantis and Border, 2006).

Theorem A.17   *Let X and Y be compact metric spaces and $\Phi : X \rightrightarrows Y$ be closed with nonempty values. Then there exists a measurable selection of Φ*

Given a metric space X and a σ-algebra $\mathcal{A}$, a *measure* is a function $\mu : \mathcal{A} \to \mathbb{R} \cup \{\infty\}$ such that $\mu(\emptyset) = 0$, $\mu(A) \geq 0$ for all $A \in \mathcal{A}$ and $\mu(\cup_{k=1}^{\infty} A_k) = \sum_{k=1}^{\infty} \mu(A_k)$ for each countable family $\{A_k\}_{k=1}^{\infty}$ of pairwise disjoint sets in $\mathcal{A}$. We say that μ is a *Borel measure* if the domain of μ is $\mathcal{B}(X)$. Furthermore, we say that μ is a *probability measure* if $\mu(X) = 1$.

The following result shows that any Borel probability measure on a metric space is inner regular (see Aliprantis and Border, 2006, Theorem 12.5, p. 436).

Theorem A.18   *If X is a compact metric space and μ is a Borel probability measure on X, then for all Borel measurable subsets B of X and $\varepsilon > 0$, there exists a closed set $F \subseteq X$ such that $\mu(B \backslash F) < \varepsilon$.*

Let X be a metric space and $M(X)$ denote the space of Borel probability measures on X. The space $M(X)$ is endowed with the following notion of convergence: a sequence $\{\mu_k\}_{k=1}^{\infty} \subseteq M(X)$ converges to $\mu \in M(X)$ if $\lim_k \int_X f d\mu_k = \int_X f d\mu$ for all bounded, continuous, real-valued functions f on X. An equivalent characterization of this form of convergence is given next (see Balder, 2011, Proposition 1). Recall that the boundary of a set $B \subseteq X$ is $\partial B = \text{cl}(B) \backslash \text{int}(B)$.

Theorem A.19   *Let X be a metric space, $\mu \in M(X)$ and $\{\mu_k\}_{k=1}^{\infty} \subseteq M(X)$. Then, $\lim_k \mu_k = \mu$ if and only if*

$$\liminf_k \int_B q \, d\mu_k \geq \int_B q \, d\mu$$

for every lower semicontinuous and bounded below $q : X \to \mathbb{R}$ and every $B \in B(X)$ such that $\alpha(\partial B) = 0$

Given a metric space X and $\mu \in M(X)$, the *support* of μ, denoted $\text{supp}(\mu)$, is the smallest closed subset of X with measure equal to 1. I.e. $\mu(\text{supp}(\mu)) = 1$ and $\text{supp}(\mu) \subseteq C$ for all closed subsets C of X such that $\mu(C) = 1$. The following results states that each probability measure on X can be approximated by a probability measure with finite support.

Theorem A.20   *Let X be a compact metric space. Then the set of probability measures with finite support is dense in $M(X)$.*

The following result shows that $M(X)$ is a compact metric space whenever X is also compact and metric (see Aliprantis and Border, 2006, Theorem 15.11, p. 513).

Theorem A.21   *If X is a compact metric space, then $M(X)$ is a compact metric space.*

The next two results establish properties of certain real-valued functions on $M(X)$ (see Aliprantis and Border (2006, Lemma 15.16, p. 516) for the first and Aliprantis and Border (2006, Theorem 15.5, p. 511) and Billingsley, 1999, Theorem 5.2, p. 31 for the second).

Theorem A.22   *Let X be a metric space and B a Borel measurable subset of X. Then the mapping $\mu \mapsto \mu(B)$, from $M(X)$ to $\mathbb{R}$, is Borel measurable.*

Theorem A.23   *Let X be a metric space and $f : X \to \mathbb{R}$ be bounded.*

1. *If f is upper semicontinuous, then the mapping $\mu \mapsto \int f\mathrm{d}\mu$, from $M(X)$ to $\mathbb{R}$, is upper semicontinuous.*
2. *If f is lower semicontinuous, then $\mu \mapsto \int f\mathrm{d}\mu$ is lower semicontinuous.*
3. *If $\mu(\{x \in X : f \text{ is discontinuous at } x\}) = 0$, then $\mu' \mapsto \int f\mathrm{d}\mu'$ is continuous at μ.*

Given metric spaces X and Y, $\mu \in M(X)$ and $\nu \in M(Y)$, the product measure $\mu \times \nu \in M(X \times Y)$ is the unique measure on $\mathcal{B}(X \times Y)$ that satisfies $\mu \times \nu(A \times B) = \mu(A)\nu(B)$ for all $A \in \mathcal{B}(X)$ and $B \in \mathcal{B}(Y)$. The following result asserts that the product measure depends continuously of each of the products (see Aliprantis, Glycopantis, and Puzzello, 2006, Lemma 3.4).

Theorem A.24   *Let $h : \prod_{i=1}^{n} M(X_i) \to M(X)$ be defined by $h(m_1, \ldots, m_n) = m_1 \times \cdots \times m_n$. Then h is continuous.*

Given metric spaces X and Y, $\mu \in M(X)$ and a function $f : X \to Y$, there is an unique measure $\pi \in M(X \times Y)$ satisfying $\pi(A \times B) = \mu(A \cap f^{-1}(B))$ for all $A \in \mathcal{B}(X)$ and $B \in \mathcal{B}(Y)$. The next result states two properties of such measure (see Balder, 2011, Propositions 2 and 3).

Theorem A.25   *Let X and Y be compact metric spaces, $f : X \to Y$ be measurable, $\mu \in M(X)$ and $\pi \in M(X \times Y)$ be such that $\pi(A \times B) = \mu(A \cap f^{-1}(B))$ for all $A \in \mathcal{B}(X)$ and $B \in \mathcal{B}(Y)$. Then:*

1. *If $\{\mu_k\}_{k=1}^{\infty} \subseteq M(X)$ is such that $\lim_k \mu_k = \mu$, then $\operatorname{supp}(\mu) \subseteq \operatorname{Ls}(\operatorname{supp}(\mu_k))$.*
2. *$\operatorname{supp}(\pi) \subseteq \operatorname{cl}(\operatorname{graph}(f))$.*

Let X and Y be metric spaces. A *transition probability* is a function $\delta : X \to M(Y)$ such that $x \mapsto \delta(x)(B)$ is measurable for all $B \in \mathcal{B}(Y)$. Given metric spaces X and Y, and $\pi \in M(X \times Y)$, then, $\pi_X \in M(X)$ denotes the marginal of π on X and is defined by $\pi_X(B) = \pi(B \times Y)$ for all $B \in \mathcal{B}(X)$. The marginal of π on Y, π_Y, is defined in an analogous way.

The following result shows how to define a measure in $M(X \times Y)$ from a measure on X and a transition probability $\delta : X \to M(Y)$ and that a Fubini-type result holds (see Neveu, 1965, Proposition III.2.1).

Theorem A.26   *Let X and Y be compact metric spaces, $\mu \in M(X)$ and $\delta : X \to M(Y)$ be a transition probability. Then there exists a probability measure $\pi \in M(X \times Y)$ such that $\pi(A \times B) = \int_A \delta(x)(B)\mathrm{d}\pi_X(x)$ for all $A \in \mathcal{B}(X)$ and $B \in \mathcal{B}(Y)$.*

Furthermore, for all bounded measurable functions $f : X \times Y \to \mathbb{R}$, the function $x \mapsto \int_Y f(x,y)\mathrm{d}\delta(x)(y)$ is measurable for all $x \in X$ and

$$\int_{X \times Y} f(x,y)\mathrm{d}\pi(x,y) = \int_X \left(\int_Y f(x,y)\mathrm{d}\delta(x)(y) \right) \mathrm{d}\mu(x).$$

The following disintegration result states that all measures in $M(X \times Y)$ are defined by integrating some transition probability (see Valadier, 1973).

Theorem A.27   *Let X and Y be compact metric spaces and $\pi \in M(X \times Y)$. Then there exists a transition probability $\delta : X \to M(Y)$ such that*

$$\pi(A \times B) = \int_A \delta(x)(B)\mathrm{d}\pi_X(x)$$

for all $A \in \mathcal{B}(X)$ and $B \in \mathcal{B}(Y)$.

Bibliography

Al-Najjar, N. (2008): "Large Games and the Law of Large Numbers," *Games and Economic Behavior*, 64, 1–34.

Aliprantis, C., and K. Border (2006): *Infinite Dimensional Analysis*. Springer, Berlin, 3rd edn.

Aliprantis, C., D. Glycopantis, and D. Puzzello (2006): "The Joint Continuity of the Expected Payoff Function," *Journal of Mathematical Economics*, 42, 121–130.

Bagh, A. (2010): "Variational Convergence: Approximation and Existence of Equilibrium in Discontinuous Games," *Journal of Economic Theory*, 145, 1244–1268.

Bagh, A., and A. Jofre (2006): "Reciprocal Upper Semicontinuity and Better Reply Secure Games: A Comment," *Econometrica*, 74, 1715–1721.

Balder, E. (2002): "A Unifying Pair of Cournot-Nash Equilibrium Existence Results," *Journal of Economic Theory*, 102, 437–470.

—— (2011): "An Equilibrium Closure Result for Discontinuous Games," *Economic Theory*, 48, 47–65.

Barelli, P., and I. Soza (2010): "On the Existence of Nash Equilibria in Discontinuous and Qualitative Games," University of Rochester.

Baye, M., G. Tian, and J. Zhou (1993): "Characterizations of the Existence of Equilibria in Games with Discontinuous and Non-quasiconcave Payoffs," *Review of Economics Studies*, 60, 935–948.

Bich, P. (2009): "Existence of pure Nash Equilibria in Discontinuous and Non Quasiconcave Games," *International Journal of Game Theory*, 38, 395–410.

Billingsley, P. (1999): *Convergence of Probability Measures*. Wiley, New York, 2nd edn.

Browder, F. (1968): "The Fixed Point Theory of Multi-valued Mappings in Topological Vector Spaces," *Mathematische Annalen*, 177, 283–301.

Carbonell-Nicolau, O. (2010): "Essential Equilibria in Normal-Form Games," *Journal of Economic Theory*, 145, 421–431.

Carbonell-Nicolau, O., and R. McLean (2011): "Approximation Results for Discontinuous Games with an Application to Equilibrium Refinement," *Economic Theory*, forthcoming.

Carbonell-Nicolau, O., and E. Ok (2007): "Voting over Income Taxation," *Journal of Economic Theory*, 134, 249–286.

Carmona, G. (2005): "On the Existence of Equilibria in Discontinuous Games: Three Counterexamples," *International Journal of Game Theory*, 33, 181–187.

—— (2009): "An Existence Result for Discontinuous Games," *Journal of Economic Theory*, 144, 1333–1340.

—— (2010): "Polytopes and the Existence of Approximate Equilibria in Discontinuous Games," *Games and Economic Behavior*, 68, 381–388.

—— (2011a): "Reducible Equilibrium Properties: Comments on Recent Existence Results," University of Cambridge.

—— (2011b): "Symposium on: Existence of Nash Equilibria in Discontinuous Games," *Economic Theory*, 48, 1–4.

—— (2011c): "Understanding Some Recent Existence Results for Discontinuous Games," *Economic Theory*, 48, 31–45.

Carmona, G., and J. Fajardo (2009): "Existence of Equilibrium in the Common Agency Model with Adverse Selection," *Games and Economic Behavior*, 66, 749–760.

Carmona, G., and K. Podczeck (2009): "On the Existence of Pure Strategy Nash Equilibria in Large Games," *Journal of Economic Theory*, 144, 1300–1319.

—— (2010): "On the Existence of Equilibria in Discontinuous Games with a Continuum of Players," University of Cambridge and Universität Wien.

—— (2011): "On the Relationship between the Existence Results of Reny and of Simon and Zame for Discontinuous Games," University of Cambridge and Universität Wien.

—— (2012): "Approximation and Characterization of Nash Equilibria of Large Games," University of Cambridge and Universität Wien.

Cauty, R. (2001): "Solution du Problème de Point Fixe de Schauder," *Fundamenta Mathematicae*, 170, 231–246.

Dasgupta, P., and E. Maskin (1986a): "The Existence of Equilibrium in Discontinuous Economic Games, I: Theory," *Review of Economics Studies*, 53, 1–26.

—— (1986b): "The Existence of Equilibrium in Discontinuous Economic Games, II: Applications," *Review of Economics Studies*, 53, 27–??.

de Castro, L. (2011): "Equilibria Existence in Regular Discontinuous Games," *Economic Theory*, 48, 67–85.

Fort, M. K. (1951): "Points of Continuity of Semi-Continuous Functions," *Publicationes Mathematicae Debrecen*, 2, 100–102.

Harris, C., P. Reny, and A. Robson (1995): "The Existence of Subgame-Perfect Equilibrium in Continuous Games with Almost Perfect Information: A Case for Public Randomization," *Econometrica*, 63, 507–544.

Harsanyi, J. (1973): "Oddness of the Number of Equilibrium Points: A New Proof," *International Journal of Game Theory*, 2, 235–250.

Jackson, M., and J. Swinkels (2005): "Existence of Equilibrium in Single and Double Private Value Auctions," *Econometrica*, 73, 93–139.

Khan, M., K. Rath, and Y. Sun (1997): "On the Existence of Pure Strategy Equilibria in Games with a Continuum of Players," *Journal of Economic Theory*, 76, 13–46.

Khan, M., and Y. Sun (1999): "Non-Cooperative Games on Hyperfinite Loeb Spaces," *Journal of Mathematical Economics*, 31, 455–492.

—— (2002): "Non-Cooperative Games with Many Players," in *Handbook of Game Theory, Volume 3*, ed. by R. Aumann and S. Hart. Elsevier, Holland.

Lebrun, B. (1996): "Existence of an Equilibrium in First Price Auctions," *Economic Theory*, 7, 421–443.

Lucchetti, R. and F. Patrone (1986): "Closure and Upper Semicontinuity Results in Mathematical Programming, Nash and Economic Equilibria," *Optimization*, 17, 619–628.

Mas-Colell, A. (1984): "On a Theorem by Schmeidler," *Journal of Mathematical Economics*, 13, 201–206.

McLennan, A., P. Monteiro, and R. Tourky (2011): "Games with Discontinuous Payoffs: A Strengthening of Reny's Existence Theorem," *Econometrica*, 79, 1643–1664.

Monteiro, P., and F. Page (2007): "Uniform Payoff Security and Nash Equilibrium in Compact Games," *Journal of Economic Theory*, 134, 566–575.

Nash, J. (1950): "Equilibrium Points in N-person Games," *Proceedings of the National Academy of Sciences*, 36, 48–49.

Nessah, R. (2011): "Generalized Weak Transfer Continuity and Nash Equilibrium," *Journal of Mathematical Economics*, 47, 659–662.

Neveu, J. (1965): *Mathematical Foundations of the Calculus of Probability*. Holden-Day, San Francisco.

Prokopovych, P. (2011): "On Equilibrium Existence in Payoff Secure Games," *Economic Theory*, 48, 5–16.

Radzik, T. (1991): "Pure-Strategy ε-Nash Equilibrium in Two-Person Non-zero-Sum Games," *Games and Economic Behavior*, 3, 356–367.

Rashid, S. (1983): "Equilibrium Points of Non-atomic Games: Asymptotic Results," *Economics Letters*, 12, 7–10.

Rath, K. (1992): "A Direct Proof of The Existence of Pure Strategy Equilibria in Games with a Continuum of Players," *Economic Theory*, 2, 427–433.

Reny, P. (1996): "Local Payoff Security and the Existence of Pure and Mixed Strategy Equilibria in Discontinuous Games," University of Pittsburgh.

—— (1999): "On the Existence of Pure and Mixed Strategy Equilibria in Discontinuous Games," *Econometrica*, 67, 1029–1056.

—— (2009): "Further Results on the Existence of Nash Equilibria in Discontinuous Games," University of Chicago.

—— (2011a): "Nash Equilibrium in Discontinuous Games," University of Chicago.

—— (2011b): "Strategic Approximations of Discontinuous Games," *Economic Theory*, 48, 17–29.

Schmeidler, D. (1973): "Equilibrium Points of Nonatomic Games," *Journal of Statistical Physics*, 4, 295–300.

Simon, L. (1987): "Games with Discontinuous Payoffs," *Review of Economic Studies*, 54, 569–597.

Simon, L., and W. Zame (1990): "Discontinuous Games and Endogenous Sharing Rules," *Econometrica*, 58, 861–872.

Sion, M., and P. Wolfe (1957): "On a Game Without a Value," in *Contributions to the Theory of Games, volume III*, ed. by A. W. T. M. Dresher and P. Wolfe. Princeton University Press, Princeton.

Valadier, M. (1973): "Désintégration d'une Mesure sur un Produit," *Comptes Rendus de L'Académie des Sciences*, 276, 33–35.

van Damme, E. (1991): *Perfection and Stability of Nash equilibrium*. Springer Verlag, Berlin.

Ziad, A. (1997): "Pure-Strategy ε-Nash Equilibrium in *n*-Person Nonzero-Sum Discontinuous Games," *Games and Economic Behavior*, 20, 238–249.

Index